Henning Berghäuser

Investigation of Dalitz decays

Henning Berghäuser

Investigation of Dalitz decays

Determination of the electromagnetic transition form factors of the eta and the pion meson

Südwestdeutscher Verlag für
Hochschulschriften

Imprint
Any brand names and product names mentioned in this book are subject to trademark, brand or patent protection and are trademarks or registered trademarks of their respective holders. The use of brand names, product names, common names, trade names, product descriptions etc. even without a particular marking in this work is in no way to be construed to mean that such names may be regarded as unrestricted in respect of trademark and brand protection legislation and could thus be used by anyone.

Publisher:
Südwestdeutscher Verlag für Hochschulschriften
is a trademark of
Dodo Books Indian Ocean Ltd., member of the OmniScriptum S.R.L Publishing group
str. A.Russo 15, of. 61, Chisinau-2068, Republic of Moldova Europe
Printed at: see last page
ISBN: 978-3-8381-2431-5

Zugl. / Approved by: Gießen, UNI, Diss., 2010

Copyright © Henning Berghäuser
Copyright © 2011 Dodo Books Indian Ocean Ltd., member of the OmniScriptum S.R.L Publishing group

Dissertation zu dem Thema

Investigation of the Dalitz decays and the electromagnetic form factors of the η and π^0-meson

vorgelegt dem
Fachbereich 07 der Justus-Liebig-Universität Gießen
von

Henning Berghäuser
aus

Weilburg

II. Physikalisches Institut

Betreuer: Prof. Dr. Volker Metag

Zusammenfassung

Mit dem CB/TAPS-Experiment am Elektronenbeschleuniger MAMI-C in Mainz werden Zerfälle von Mesonen und Baryonen untersucht. Die Hadronen werden in photoninduzierten Reaktionen an Protonen (und Neutronen) produziert. Dabei können als Target nicht nur freie Protonen (lH_2) sondern auch Atomkerne (C, Nb, Pb) verwendet werden.

In der vorliegenden Arbeit wurden die Dalitz Zerfälle des η und des π^0 Mesons analysiert. Der Fokus lag dabei auf der Bestimmung des elektromagnetischen Übergangsformfaktors des η-Mesons. Dieser Formfaktor enthält wesentliche Informationen über die elektromagnetischen Eigenschaften und Substrukturen des Teilchens. Mesonen sind keine punktförmigen Elementarteilchen, sondern bestehen aus stark wechselwirkenden Quarks und Antiquarks. Demzufolge sind für die elektromagnetischen Zerfälle eines Mesons Abweichungen von den Vorhersagen der Quanten Elektrodynamik (QED) zu erwarten. Für diese Abweichungen existieren Vorhersagen unterschiedlicher theoretischer Modelle, wie z.B. dem Vector Meson Dominance Model (VMD) oder dem Leupold-Terschluesen Modell [46]. Anhand von Experimenten wie in dieser Arbeit können die theoretischen Modelle überprüft werden.
Neben den Dalitz Zerfällen wurden in der vorliegenden Arbeit weitere Zerfallskanäle der Mesonen π^0, η und ω untersucht mit dem Ziel, die entsprechenden Verzweigungsverhältnisse zu bestimmen. Darüber hinaus wurde der Wirkungsquerschnitt für die $\pi^0\eta$-Produktion am Proton bestimmt.

Ein weiterer Aspekt dieser Arbeit war die Entwicklung einer Methode zur Separation von geladenen Pionen und Elektronen/Positronen im Rahmen der Teilchenidentifikation. Diese Trennung ist eine wesentliche Voraussetzung für die Analyse geladener Zerfallskanäle. Da das CB/TAPS Experiment kein Magnetfeld verwendet, erscheint diese Separation zunächst als sehr schwierig. Jedoch wird in dieser Arbeit eine Methode vorgestellt, mit der eine Separation auch ohne ein Magnetfeld erfolgreich realisiert werden kann. Werden alle Teilchen im Endzustand einer Reaktion sowie das Rückstoßproton nachgewiesen (exklusive Analyse), so können alle Informationen der vollständigen Kinematik in der Analyse ausgenutzt werden, um missidentifizierte $\pi^\pm, e^\pm$ zu entfernen. Zusätzliche Schnitte auf die Clustergrößen helfen ebenfalls diesen Untergrund weiter zu unterdrücken. Die Wahrscheinlichkeit dafür, ein $\pi^+\pi^-$-Paar als e^+e^--Paar falsch zu identifizieren ist kleiner als $3 \cdot 10^{-7}$.

Die experimentellen Daten wurden in zwei Strahlzeiten in 2007 mit den Detektoren

Tagger, Crystal Ball und TAPS aufgezeichnet. Der primäre Photonenstrahl, der über den Bremsstrahlungseffekt aus dem Elektronenstrahl von MAMI-C erzeugt wurde, traf auf ein Flüssigwasserstofftarget auf und induzierte dabei die hadronischen Reaktionen. Eine vorläufige Teilchenidentifikation wurde mit Hilfe der dEvE-Methode und der time-of-flight-Technik realisiert. Anhand der gemessenen Zeitkoinzidenzen konnten zufällige Ereignisse aussortiert werden.

Da der Umfang der experimentellen Daten ca. 2 TByte betrug, war eine Datenkompression notwendig. Für deren Verwirklichung wurde eine spezielle, NTuple-gestützte Analyseprozedur entwickelt und nur solche Ereignisse wurden ausgewertet, die bestimmte Kriterien erfüllten. Auf diese Weise konnte die Datenmenge stark reduziert werden. Neben der Kompression wurde auch die Kalibration der Daten und die vorangehende Entwicklung dazu geeigneter Methoden und benötigter Programme an der Universität Giessen durchgeführt. Auch für die Analyse der detektierten Ereignisse wurde ein neues Programm in C++ entwickelt ($AR_{HB}2v3$).

In der exklusiven Analyse des Dalitz Zerfalls $\eta \to e^+e^-\gamma$ konnten 827 Ereignisse rekonstruiert werden. Verglichen mit dem Ergebnis der SND-Kollaboration [1] ist damit die Statistik in der vorliegenden Arbeit um einen Faktor 8 größer. Für die Steigung des η-Formfaktors wurde der folgende Wert ermittelt:

$$b = \frac{dF}{dq^2}|_{q^2=0} = \Lambda^{-2} = 1.84^{+0.43}_{-0.32}\frac{1}{\text{GeV}^2}$$
$$\text{mit:} \quad \Lambda = (740 \pm 74) \text{ MeV}$$

Innerhalb der Fehler stimmt dieses Ergebnis mit den Resultaten der Experimente Lepton-G [13], NA60 [12] und SND [1] überein. Des Weiteren konnte eine gute Übereinstimmung mit dem theoretischen Modell von Leupold-Terschlüsen [46] festgestellt werden.

In der Analyse des π^0-Dalitz Zerfalls wurde wie erwartet keine Abweichung von der QED-Vorhersage gefunden.

Neben den genannten Dalitz-Zerfällen wurden auch die Reaktionen $\eta \to \gamma\gamma$, $\eta \to \gamma\gamma\pi^0$, $\eta \to 3\pi^0$, $\eta \to \pi^0\pi^+\pi^-$, $\omega \to \pi^+\pi^-\pi^0$ sowie $\omega \to \pi^0\gamma$ untersucht. In diesem Zusammenhang konnten die folgenden Verzweigungsverhältnisse bestimmt werden:

$$BR_{\eta\text{-Dalitz}} = (6.18 \pm 0.65) \cdot 10^{-3} \qquad BR_{\eta \to \pi^+\pi^-\pi^0} = (22.9 \pm 1.7)\%$$
$$BR_{\omega \to \pi^0\gamma} = (10.2 \pm 1.4)\%$$

Diese Ergebnisse entsprechen publizierten Resultaten [20]. Es ist anzumerken, dass der Fehler des in dieser Arbeit gemessenen Wertes für $BR_{\eta-\text{Dalitz}}$ kleiner ist als der entsprechende Fehler im Particle Data Booklet der Particle Data Group.

Abstract

In this thesis the Dalitz decays of the π^0, η, and ω-meson have been studied in photon induced reactions off the proton: $\gamma+p \to \pi^0+p \to e^+e^-\gamma+p$, $\gamma+p \to \eta+p \to e^+e^-\gamma+p$, and $\gamma+p \to \omega+p \to e^+e^-\pi^0+p$.

The main aim has been to determine the electromagnetic transition form factor of the η-meson. This form factor provides significant information about the electromagnetic properties of this meson. As a meson is a non-point like particle, the corresponding electromagnetic transition form factor is expected to differ from the standard **QED** prediction (**Q**uantum **E**lectro **D**ynamics). Hence, a measurement of this form factor provides the possibility to investigate this deviation from the QED and to test theoretical models like the (**V**ector **M**eson **D**ominance Model) **VMD** or the model by Leupold-Terschluesen [46].

Beside the Dalitz decays other decay modes of the η and the ω-meson were analyzed and the branching ratios of the decays $\eta \to \pi^+\pi^-\pi^0$, $\eta + p \to e^+e^-\gamma$ and $\omega \to \pi^0\gamma$ were determined. Furthermore the cross section of η-production as well as the cross section of $\pi^0\eta$-production in photon induced reactions off the proton were determined.

Another aspect of this work was to investigate the possibility of separating electrons and positrons from charged pions with the Crystal Ball and TAPS detector systems at MAMI-C in Mainz. This coupled detector-setup is very efficient in detecting photons and thus it is particularly suited for measurements of neutral decay modes of hadrons. As this setup does not use a magnetic field, an accurate separation of e^+, e^- from π^+, π^- is difficult. However, it was shown in this work, that an accurate separation and identification of those particles is possible by exploiting the full kinematic information available in exclusive analyses. Thereby Dalitz decays were identified. The background from charged pions was suppressed further. Moreover it was found that cuts on the cluster sizes of the charged hits further suppress the $\pi^\pm$-background. The probability for the misidentification of a $\pi^+\pi^-$-pair as an e^+e^--pair is less than $3 \cdot 10^{-7}$.

The experimental data were taken during two beamtimes in 2007 at the electron acceleration facility MAMI-C in Mainz. Energy tagged photons produced via the bremsstrahlung process impinged on a liquid hydrogen target and induced among others the reactions of interest. The detectors Tagger, Crystal Ball (including the PID) and TAPS (including the VETO) were used for data recording. A preliminary particle identification was provided by the *dE-versus-E* and the *time-of-flight* method. Furthermore the information about *time-coincidences* between detected hits were exploited in order to suppress random events.

The total amount of experimental raw data was approximately 2 TByte. Therefore a compression of the data was necessary. For this purpose a ntuple-based analysis procedure was developed. Only events fulfilling certain requirements were saved to these ntuples. Thus the original amount of data was significantly reduced.

Before this compression, the data had been calibrated by the A2-group of the University of Giessen. In this respect many procedures, macros and programs had to be developed from scratch. The same holds for the subsequent analysis of the data. A new analysis program called $AR_{HB}2v3^1$ was developed in C++.

In all analyses the detection of the meson and the recoiling proton was required; thus the full kinematic information could be exploited. Cuts were applied on the energy balance, momentum balance, missing mass and the coplanarity. Depending on the particular decay channel further cuts were applied on the relative angle between particles, the incident energy, the θ-angle of the proton and if applicable on the cluster sizes of the charged hits. The applied cuts were verified by displaying each variable under the constraint of all other cuts. Besides the analysis of experimental data simulated data were analyzed in order to determine the detector response. The simulated data were produced by a Monte-Carlo simulation which contained the full detector setup. The corresponding start distributions for the simulation were generated using a phase space event generator as well as the PLUTO event generator.

In the exclusive analysis of $\eta \to e^+e^-\gamma$ 827 events were reconstructed. This is an improvement by a factor 8 with respect to the measurement by the SND collaboration [1]. The slope parameter of the associated transition form factor was determined as

$$b = \frac{dF}{dq^2}|_{q^2=0} = \Lambda^{-2} = 1.84^{+0.43}_{-0.32}\frac{1}{\text{GeV}^2}$$
$$\text{with:} \quad \Lambda = (740 \pm 74) \text{ MeV}$$

Within the errors this result is consistent with the result of the Lepton-G experiment [13], the NA60 experiment [12], and the SND experiment [1]. Furthermore the result agrees within the errors with the theoretical prediction of [46].

In the investigation of the π^0-Dalitz analysis no deviation from the QED-prediction was found. In the case of the ω-Dalitz decay a from factor could not be determined because of limited statistics.

The determined branching ratios are:

$$BR_{\eta\text{-Dalitz}} = (6.18 \pm 0.65) \cdot 10^{-3} \qquad BR_{\eta \to \pi^+\pi^-\pi^0} = (22.9 \pm 1.7)\%$$
$$BR_{\omega \to \pi^0\gamma} = (10.2 \pm 1.4)\%$$

These results are consistent with the values in the Particle Data Booklet of the Particle Data Group. Furthermore the obtained cross sections for η-production and for $\pi^0\eta$-production (Chapter 7) are in agreement with the results of former publications [26].

---

[1]This program is based on the AcquRoot-4v2-System from J.R.R. Annand, University of Glasgow, which provides all the data-decoding functionality but only rudimentary analysis procedures.

Contents

Zusammenfassung	III
Abstract	V

1. Introduction — 1
  1.1. About this thesis 1
      1.1.1. Structure of this thesis 2
      1.1.2. Used units 3
      1.1.3. The labeling of the axes of histograms 3
  1.2. Theory 3
      1.2.1. The standard model 3
      1.2.2. Hadrons 6
      1.2.3. The Masses of Hadrons 12
      1.2.4. Electromagnetic form factors 14
      1.2.5. The Vector Meson Dominance Model 18
  1.3. Former Experiments and Results 20

2. The experimental setup — 23
  2.1. The MAMI accelerator facility 23
  2.2. The Glasgow Photon Tagging Spectrometer 25
      2.2.1. Photon Beam and Tagging Efficiency Measurement 28
  2.3. The Liquid Hydrogen Target 29
      2.3.1. Probability for a $\gamma \rightarrow e^+e^-$-conversion in the target region 30
      2.3.2. Number of target protons 31
      2.3.3. Other targets 31
  2.4. Crystal Ball System 33
      2.4.1. The NaI Calorimeter 33
      2.4.2. The Particle Identification Detector 35
      2.4.3. The Multi-Wire Proportional Chambers 37
  2.5. The TAPS Detector 37
      2.5.1. The $BaF2$-Calorimeter 39
      2.5.2. The $PbWO_4$-Upgrade 40
      2.5.3. The VETO Wall 41
  2.6. Electronics, data acquisition and read-out 43
      2.6.1. Crystal Ball 43

	2.6.2. TAPS electronics	45
2.7.	The Trigger	47
2.8.	Beam-time overview	48

3. Calibration   51
- 3.1. NaI Energy Calibration 51
  - 3.1.1. Basic Calibration 52
  - 3.1.2. Linear Calibration 53
  - 3.1.3. Second order Calibration 56
- 3.2. NaI Time Calibration 59
  - 3.2.1. CB Time Walk Correction 60
- 3.3. BaF_2 Energy Calibration 62
  - 3.3.1. Cosmic Calibration 63
  - 3.3.2. Linear energy Calibration 64
  - 3.3.3. Second order Calibration 65
- 3.4. BaF_2 Time Calibration 66
- 3.5. PID ϕ Correlation 69
- 3.6. PID Energy Calibration 70
- 3.7. VETO Energy Calibration 72
- 3.8. VETO Correlation and Time Calibration 73
- 3.9. TAGGER Energy Calculation 74
- 3.10. TAGGER Time Calibration 75
- 3.11. Readout of the TAGGER Scalers 75
- 3.12. The Tagging Efficiency Measurement 77
- 3.13. The Photon Flux 78
- 3.14. Verification of the Energy Calibration 80

4. Particle identification and event reconstruction   83
- 4.1. Software 83
  - 4.1.1. The AcquRoot Analyzer 83
  - 4.1.2. The HBAnalysis1v8 NTuple Analyzer 89
  - 4.1.3. The Event Summary Data Project 91
  - 4.1.4. The A2-Sim Monte Carlo Simulation 92
- 4.2. Hardware 96
- 4.3. Analysis Procedure 98
  - 4.3.1. Particle Reconstruction in CB 99
  - 4.3.2. Particle Reconstruction in TAPS 100
  - 4.3.3. Reconstruction and Separation of e^+e^- from $\pi^+\pi^-$ 101
  - 4.3.4. Event Selection and Data Compression 105
  - 4.3.5. Application of Cuts 106

5. Simulation   107
- 5.1. Two Body Calculations 107

		5.1.1. η-Production Calculations 107

      5.1.1. η-Production Calculations 107
      5.1.2. ω-Production Calculations 108
      5.1.3. π^0-Production Calculations 110
      5.1.4. η'-Production Calculations 111
  5.2. Start Distributions of the Dalitz simulations 111
      5.2.1. η-Dalitz Phase Space Distribution 112
      5.2.2. η-Dalitz PLUTO Distribution 112
      5.2.3. ω-Dalitz PLUTO Distribution 113
      5.2.4. π^0-Dalitz PLUTO Distribution 114
      5.2.5. Comparison between Start distributions: PLUTO vs. Phase Space 115
  5.3. Energy scaling 122

6. Analysis   125

  6.1. Analysis of simulated data 125
      6.1.1. SIM: Exclusive analysis of $\eta \to \gamma\gamma$ 126
      6.1.2. SIM: Exclusive analysis of $\eta \to \pi^0\pi^0\pi^0$ 129
      6.1.3. SIM: Exclusive analysis of $\eta \to e^+e^-\gamma$ 129
      6.1.4. SIM: Exclusive analysis of $\eta \to \pi^+\pi^-\gamma$ 133
      6.1.5. SIM: Exclusive analysis of $\eta \to \pi^+\pi^-\pi^0$ 136
      6.1.6. SIM: Exclusive analysis of $\omega \to \pi^0\gamma$ 137
      6.1.7. SIM: Exclusive analysis of $\omega \to \pi^+\pi^-\pi^0$ 140
      6.1.8. SIM: Exclusive analysis of $\omega \to \pi^0 e^+e^-$ 140
      6.1.9. SIM: Exclusive analysis of $\pi^0\eta$-production 141
      6.1.10. SIM: Exclusive analysis of $\pi^0 \to e^+e^-\gamma$ 143
  6.2. Analysis of experimental data 145
      6.2.1. DATA: Exclusive analysis of $\eta \to \gamma\gamma$ 147
      6.2.2. DATA: Exclusive analysis of $\eta \to \pi^0\pi^0\pi^0$ 148
      6.2.3. DATA: Exclusive analysis of $\eta \to \pi^0\gamma\gamma$ 149
      6.2.4. DATA: Exclusive analysis of $\eta \to e^+e^-\gamma$ 150
      6.2.5. DATA: Exclusive analysis of $\eta \to \pi^+\pi^-\pi^0$ 162
      6.2.6. DATA: Exclusive analysis of $\omega \to \pi^0\gamma$ 162
      6.2.7. DATA: Exclusive analysis of $\omega \to \pi^+\pi^-\pi^0$ 166
      6.2.8. DATA: Exclusive analysis of $\omega \to \pi^0 e^+e^-$ 167
      6.2.9. DATA: Exclusive analysis of $\pi^0\eta$-production 169
      6.2.10. DATA: Exclusive analysis of $\pi^0\pi^0$-production 172
      6.2.11. DATA: Exclusive analysis of $\pi^0 \to e^+e^-\gamma$ 174
  6.3. Determination of branching ratios 176
      6.3.1. Branching Ratio of $\eta \to e^+e^-\gamma$ 179
      6.3.2. Branching Ratio of $\eta \to \pi^+\pi^-\pi^0$ 179
      6.3.3. Branching Ratio of $\omega \to \pi^0\gamma$ 179
  6.4. Discussion of Background channels 180
      6.4.1. Background channels in the analysis of $\eta \to e^+e^-\gamma$ 180
      6.4.2. Background channels in the analysis of $\omega \to \pi^0\gamma$ 194

7. Results		**197**
7.1.	The measured channels	197
7.2.	The separation of e^+e^- from $\pi^+\pi^-$	199
7.3.	The η-Dalitz decay	200
	7.3.1. The transition form factor	202
7.4.	The Dalitz decays of the π^0-meson	203
7.5.	Conclusion and outlook	207
A. Appendix		**211**
B. List of Figures		**223**
C. List of Tables		**225**
D. Bibliography		**227**

1. Introduction

1.1. About this thesis

The aim of this thesis is the investigation of Dalitz decays of neutral mesons. The analyses of Dalitz decays are of general interest. This is because these decays provide a possibility to obtain information about the electro-magnetic properties of the mesons as well as about the substructures of these particles. The decay of any meson into e^+e^- can be calculated and predicted with the **Q**uantum **E**lectro**d**ynamics (QED). However, the QED can only deliver correct predictions for point like particles. As mesons are not point-like, a measurement of the Dalitz decays is important. As a matter of fact the experimental results should differ from the QED predictions and this has already been found in former experiments (section 1.3). The question is, whether the results of the performed analyses of experimental data are consistent with the **V**ector **M**eson **D**ominance model (VMD) or not. Although some former experiments already seem to confirm the VMD, still more confirmation is necessary.

To contribute experimental data the A2-Collaboration (CB/TAPS @ MAMI-C) performed a fixed target experiment on LH_2. The data has been obtained from June 2007 to July 2007 at the accelerator facility MAMI-C in Mainz using the combined detector systems of TAGGER, Crystal Ball, and TAPS, thereby covering the complete 4π solid angle.
Besides the production of π^0 and η also ω-mesons, which have a production threshold of 1108 MeV were produced. This was only possible due to an upgrade of the electron accelerator allowing to accelerate electrons up to 1.5 GeV. The following reactions were of main importance in the analyses of the data:

$$\pi^0 \rightarrow e^+e^-\gamma \qquad (1.1)$$
$$\eta \rightarrow e^+e^-\gamma \qquad (1.2)$$
$$\omega \rightarrow e^+e^-\pi^0 \qquad (1.3)$$

The focus in the analysis was set on the Dalitz decay of η meson. The reason for this was, that in the case of the π^0 Dalitz decay no VMD effect is assumed to be observable. Concerning the ω-meson, the statistics was assumed not to be high enough to match the requirement for a significant result.

1. Introduction

Still, the ω Dalitz as well as the π^0 Dalitz decay were investigated. Besides this, other charged decay modes of the η and the ω meson were analyzed and furthermore some neutral decay modes were investigated too:

$$
\begin{aligned}
\text{Other charged decays} &: \\
\eta &\to \pi^+\pi^-\pi^0 \\
\omega &\to \pi^+\pi^-\pi^0 \\
\text{Neutral decays and reactions} &: \\
\pi^0 &\to \gamma\gamma \\
\eta &\to \gamma\gamma \\
\eta &\to \pi^0\gamma\gamma \\
\eta &\to \pi^0\pi^0\pi^0 \\
\omega &\to \pi^0\gamma \\
\pi^0\pi^0 &- \text{Production} \\
\pi^0\eta &- \text{Production}
\end{aligned}
\quad (1.4)
$$

1.1.1. Structure of this thesis

In this chapter the basic concepts of elementary physics will be discussed as well as the theoretical framework behind it. In this context the standard model, which describes the systematics of elementary particles, will be discussed. Furthermore the properties of the η-meson and the ω-meson will be listed, as these mesons are of essential importance to this work. Concerning the ω-meson in-medium effects will be discussed, because these investigations are one of the major activities of the A2-analysis group at the University of Giessen.
Furthermore a closer look on the electromagnetic transition form factors as well as on the **V**ector **M**eson **D**ominance model (**VMD**) will be given, as the main topic of this investigation is the analysis of the Dalitz decays of neutral mesons (π^0, η, ω).

Chapter 2 explains the experimental setup of CB/TAPS at the electron accelerator facility MAMI-C in Mainz. All components of the detectors TAGGER, TAPS and Crystal Ball will be described.

The following chapter (3) describes the calibration procedures; detailed information about every step is given.

In chapter 4 the developed and used software is presented. Furthermore a detailed introduction into the procedure of particle/event reconstruction is provided and the complete chain of analysis is explained.

The next chapter (5) contains information about all simulations accomplished by the author.

In chapter 6 all accomplished analyses are described and the result of each investigation will be presented.

In a final step the results obtained are presented. In chapter 7 these results are discussed and compared to available theoretical predictions.

1.1.2. Used units

The common units in the field of 'particle physics' and 'high energy physics' will be used in this thesis. The unit of energy is the *electron Volt* (eV), which is the energy that a particle with one unit of charge (e.g. an electron) acquires when it passes through a potential of 1 Volt. 1 eV is equivalent to $1.602 \cdot 10^{-19}$ Joule. The equations $E = m \cdot c^2$ and $E = p \cdot c$ imply that $\frac{eV}{c^2}$ is the unit for mass and $\frac{eV}{c}$ the unit for the momentum. Thereby c is the speed of light and a very well known constant. In this thesis the so-called *natural unit system* is used, which sets $c = 1$ and $h = 1$. The latter is the Planck constant. Thus the electron Volt is the given unit for mass, energy and momentum.

1.1.3. The labeling of the axes of histograms

As with ROOT version 5.22 the usage of the Latex module often led to a crush of the ROOT-CINT while editing histograms, the decision was made not to use Latex font setting for the labeling of the axis of histograms. Thus, in many histograms $\gamma\gamma$ will be written as gg and e^+e^- will be displayed as $e+e-$. Furthermore $\pi^{\pm,0}$ will be written as *pi+*, *pi-*, *pi0* and *eta* will be used for the η.

1.2. Theory

1.2.1. The standard model

The so-called **S**tandard **M**odel of particle physics describes successfully all known particles and effective forces between these particles, based on what we know today. During the last decades many predictions of this model could be proofed to be valid.

The standard model contains the elementary particles (Table 1.1), which are the basic components of all other particles. In total 60 elementary particles have been discovered

1. Introduction

in nature and are included in the standard model. These are 6 leptons, the corresponding 6 anti-leptons, 6 types of quarks (each can wear one out of three different colors), the corresponding 6 anti-quarks and 12 bosons (8 gluons (g), the W^+, W^-, Z^0 and the photon γ); see Table 1.2.

Fermion	Family			Charge [e]	Color-Charge
Leptons	e	μ	τ	-1	-
	ν_e	ν_μ	ν_τ	0	-
Quarks	u	c	t	$+2/3$	$r, g, b, \bar{r}, \bar{g}, \bar{b}$
	d	s	b	$-1/3$	$r, g, b, \bar{r}, \bar{g}, \bar{b}$

Table 1.1.: The elementary particles contained in the standard model. Leptons and Quarks are Fermions, meaning that their Spin is equal to 1/2. Not listed here but also included in the Standard Model are the corresponding anti-particles [44].

Force	Coupling-Mechanism	Field Boson
strong	color charge	8 Gluons (g)
electromagnetic	electric charge	Photon γ
weak	weak charge	$W^{+/-}, Z^0$

Table 1.2.: Elementary forces of the Standard Model and their field bosons [44].

- **Electroweak force:**
  In 1967 A. Salam, S. Glashow and S. Weinberg introduced a formalism in which the weak and the electromagnetic force were combined and could be treated as two aspects of one unified interaction. This formalism introduced the weak isospin as a new quantum number and apart from that it is built up analogously to the Isospin formalism of the strong interaction, which will be described further below. As we now already know, the electroweak force consists of two parts:

  1. **Electromagnetic force:**
     The **Q**uantum **E**lectro **D**ynamics(QED) describes all electromagnetic processes properly via the exchanges of photons (as field bosons) and has meanwhile become well established. Moreover the QED has advanced to the most precise theory in science. Important facts are that the exchange boson is mass-less and charge-less. Furthermore the range of the electromagnetic force is infinite. That is, because the interaction range is given by the Compton radius, which is defined as $R = 1/M$, with M being the mass of the exchange boson, which is in this case the mass-less photon.

  2. **Weak fore:**
     The weak force affects all particles but it only plays a role in those interactions in which neither the electromagnetic force nor the strong force contribute.

1.2. Theory

This fact can be explained by the relative weakness compared to the other two forces. A well known example for a weak interaction is the radioactive β-decay. The field bosons of the weak force are the W^+, W^- and the Z^0 bosons. They are very massive (Table 1.3) and thus the range of the interaction has to be very small (which can be explained again by the Compton radius).

In the electroweak unification the three bosons differ in their third Isospin component (assuming this component is conserved in reactions with charged currents). The W^- boson has $T_3(W^-) = -1$, and the W^+ has $T_3(W^+) = +1$. Further on the W^0 with $T_3(W^0) = 0$ completes the Isospin triplet, whereas W^0 is not the Z^0, which is the field boson of the uncharged weak interaction. Experiments performed at LEP and SLAC verified this theory. Although the bosons forming this tripled should couple with equal strength to all members of a multiplet[1], this could not be observed in the experiments. In contrast, it was found that the decay probability of the Z^0 into charged leptons is less than into neutrinos. This fact rules out the possibility of $W^0 = Z^0$. As a solution another state called B^0 has to be postulated, which is a weak Isospin singlet with $T = 0$ and $T_3 = 0$. Further on the two experimentally observed neutral vector bosons (photon and Z^0) are described as orthogonal linear combinations of the W^0 and the B^0. The mixture of these two states is characterized by the so-called *Weinberg angle* Θ_W:

$$|\gamma\rangle = \cos\Theta_W |B^0\rangle + \sin\Theta_W |W^0\rangle \quad (1.5)$$
$$|Z^0\rangle = -\sin\Theta_W |B^0\rangle + \sin\Theta_W |W^0\rangle \quad (1.6)$$

Within this theory the absolute mass of the Z^0 and $W^\pm$ bosons were predicted before these bosons were observed. The electroweak unification was honored with the Nobel prize for A. Salam, S. Glashow and S. Weinberg in 1979.

Still one problem of the electroweak unification remains to be solved. A mixture should only occur when the states have similar energies (masses). As the γ is massless but the Z^0 has a very high mass (Table 1.3) a mixture should not occur. In order to solve this problem other theoretical models were invented. The most prominent model, is the Higgs mechanism introduced by Higgs [23]. It is based on the concept of phase transitions and the postulation of a spontaneous symmetry breaking with an unsymmetric ground state. This is a well established concept in physics (e.g. ferro magnetism). However, in the Higgs model, four *Higgs fields* are postulated. The photon, Z^0 and the $W^\pm$ are massless above the phase transition at a certain energy. Below this energy their masses are generated through three out of the four *higgs fields*. Since the photon remains massless, the fourth *higgs field* does not get absorbed and thus should be observable in nature. As the statistics of former CERN experiments was insufficient and did not allow conclusive statements about some detected candidates for the *Higgs-boson*, it is up to the next generation experiments to observe and identify the *Higgs*. This task was one of the major motivations for building the **L**arge **H**adron **C**ollider (LHC) at CERN.

---

[1] e.g to electrons and neutrons which form a doublet (e^+, ν_e).

1. Introduction

Interaction	strong	electromagnetic	weak
relative strength	1	10^{-2}	10^{-6}
typical decay time	$\sim 10^{-23} s$	$\sim 10^{-26} - 10^{-19} s$	$\sim 10^{-8}$
exchange boson	8 gluons	photon (γ)	W^+, W^-, Z^0
mass	0	0	$W^\pm \approx 80$ GeV $Z^0 \approx 90$ GeV

Table 1.3.: Properties of elementary forces.

- **Strong force:** As indicated by the name the strong force is the strongest of the elementary forces. Different is, that this force is limited to very small distances in the order of fm^2. This is roughly the diameter of a nucleon and helps to explain, why protons in a nucleus do not push each other away; it simply over-compensates the electromagnetic repulsion in between protons. Hence, the strength of the strong force is the reason for stable nuclei. The field theory that describes the strongly interacting particles is called **Quantum Chromo Dynamics** (**QCD**). In QCD the color-charge[3] is the source of the interaction and 8 gluons are the field bosons. As gluons carry color themselves, they can couple to each other. This is a major difference to QED, in which the field boson is the neutral γ, that does not carry any charge itself; and thus it can not couple to other photons. The fact that gluons can interact among each other is the reason for the very short interaction range of the strong force.

- **Gravitation:**
Gravitation is the longest known of all forces and it affects all particles depending on their masses. Until today it is the least understood in the sense of a quantum field theoretical description. The exchange boson has been postulated but has not yet been discovered, nor the classical analogon, gravitational waves. Fortunately, the gravitation does not affect elementary particles very much, at least only in such a manner, that interactions based on the other forces are still dominating and the results of corresponding investigations are not distorted. The reason for this is the very small mass of each of the elementary particles. Hence, this force can be discarded when dealing with basic interactions of elementary particles and thus it is not (yet) included in the standard model.

1.2.2. Hadrons

All strongly interacting particles that are observable in nature are composite systems of quarks and anti-quarks as constituents; these are called *hadrons*. A hadron that consists

---
[2] fm stands for femto meter: $1 fm = 10^{-15}$m.
[3] Originally, the concept of color charge was only introduced for pure theoretical reasons, namely in order to obey the Pauli principle. Nowadays plenty of experimental results support this concept.

of three quarks is called a baryon, and a particle consisting of one quark and one antiquark is called a meson. Mesons and baryons have in common, that they are colorless. The reason for this is the *confinement*. So far, free colored states seem not to exist, as all experiments have failed to observe a free quark until today. Thus quarks (the elementary particles carrying the colors) have to combine to systems in such a manner, that the total color of the combined system becomes white (colorless[4]). Furthermore the existence of a penta-quark has been claimed, as it is possible to combine four quarks and one antiquark in such a manner, that the resulting total color is white. Various experiments reported evidence for such a state, but the majority of the experiments does not support this claim. Moreover more complex systems built of quarks, anti-quarks and gluons, or even poor glueballs should exist. Once again, many experiments report evidence, but with low statistics, and all results are still under debate.

Baryons

As was pointed out before, baryons are qqq-states, in other words: they consist of three quarks. Due to the number of 6 available quark-flavors and the larger combinatorial possibilities, the total number of different baryons should be far greater than the one of mesons. Hence, this number makes the baryon wave functions more complex than those for the mesons.

The lightest baryons couple to a total angular momentum of $L = 0$, and the spin can either couple to $S = \frac{1}{2}$ or $S = \frac{3}{2}$. Thus, the symmetry for a qqq-state consisting only of (u,d,s)-quarks is given by:

$$3 \otimes 3 \otimes 3 = 10_s \otimes 8_m \otimes 8_m \otimes 1a \tag{1.7}$$

Whereas these are one symmetric decuplet, two octets with mixed symmetry and one antisymmetric singlet. If one now includes the two possible spin orientations, the underlying symmetry increases to:

$$6 \otimes 6 \otimes 6 = 56_s \otimes 70_m \otimes 70_m \otimes 20a \tag{1.8}$$

Further on these multiplets can be subdivided again.

$$56 = {}^4 10 + {}^2 8 \tag{1.9}$$
$$70 = {}^2 10 + {}^4 8 + {}^4 8 + {}^2 1 \tag{1.10}$$
$$20 = {}^2 8 + {}^4 1 \tag{1.11}$$

Thereby the superscript is $2S + 1$, which is equal to the amount of possible spin orientations within each multiplet. In the ground state ($L = 0$), the $J^P = 1/2^+$ octet and the $J^P = 3/2^+$ decuplet together make up 18 possible states. In the Figures 1.1 and 1.2 the

---
[4]E.g. in a meson one of the two $\bar{q}q$ has always to carry the anti-color of the color the other quark has.

1. Introduction

lightest baryon octet and the lightest decuplet are shown. The most common members of the baryon family are the proton and the neutron, which both have the quantum number $J^P = 1/2^+$ and belong to octet in Figure 1.1. More details on all baryons can be found in [19].

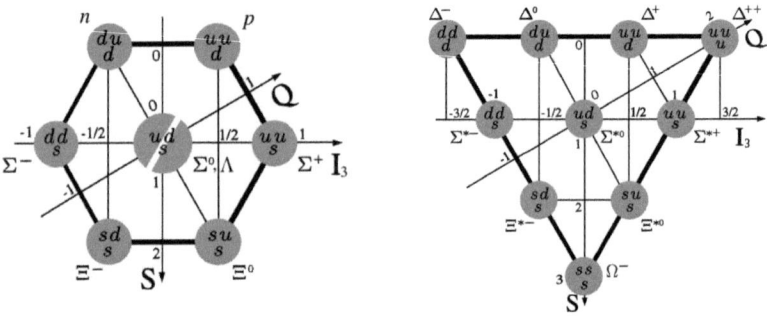

Figure 1.1.: The lightest baryon octet. The figure is taken from [52].

Figure 1.2.: The lightest baryon decuplet. The figure is taken from [53].

Mesons

Mesons are strongly interacting particles with a $q\bar{q}$ substructure. All observed mesons can be classified via their quantum numbers. This ordering principle was invented by M. Gell-Mann[5]. Focusing only on the three lightest quarks (u,d and s) the $q\bar{q}$ systems can be described using a $3 \otimes 3$ symmetry. Figure 1.3 illustrates the octet of the pseudoscalar mesons. For a better understanding of such a figure, a closer look at the quantum numbers is useful:

- Spin $S_{q\bar{q}}$:
  The spin is defined as the intrinsic angular momentum. As quarks belong to fermion-family, their spin is 1/2. Thus the orientations $-1/2$ and $+1/2$ are possible. Hence, the spin of two fermions can couple to a value between

  $$S_{q\bar{q}} = |S_q - S_{\bar{q}}| \quad \text{and} \quad S_{q\bar{q}} = |S_q + S_{\bar{q}}|$$

  As a result the total spin of a meson can either be 0 or 1.

---
[5]In 1969 M. Gell-Mann was awarded with the Nobel prize for his work.

1.2. Theory

- Orbital angular momentum L:
  Whenever two particles couple, the resulting relative angular momentum is expressed by L. For the low mass states (S-states) the orbital angular momentum couples to $L = 0$.

- The total angular momentum J:
  This is the composition of S and L, in other words: spin and orbital momentum couple to the total angular momentum of the states. The possible meson-states are:
  $$|L - S_{q\bar{q}}| \leq L \leq |L + S_{q\bar{q}}|$$
  Hence, the meson states lowest in mass have $L = 0$ and thus couple to:
  $$J = 0 \quad \text{or} \quad J = 1$$

- Parity P:
  The simultaneous flip in the sign of all spatial coordinates is called the *Parity* transformation. As parity describes the behavior of a particle under performing a spatial inversion, this quantum number can have the values $+1$ or -1.
  The total parity of a combined state is made up of all the intrinsic parities of the constituents and the parity of the binding itself. Thus parity is of multiplicative nature. The parity quantum numbers of quarks and anti-quarks are of opposite sign (a factor of -1). Hence, the parity of the spatial wave function of a meson can be determined by the angular orbital momentum and gives rise to a factor of $(-1)^L$. Thus the total parity of a mesonic state is therefore:
  $$P = (-1)^{L+1}$$

- Isospin I:
  Heisenberg[6] introduced the so-called Isospin formalism, which is based on the idea that the proton and the neutron are the same type of particles, but they can only be identified by their electric charge (0 and +1). In this formalism the basic constituents of the nucleon, the u-quark and the d-quark, are treated as different states of the same particle.
  Furthermore the Isospin formalism is applied and handled in the same mathematical way as is the conventional spin formalism; thus the third component I_3 determines the states (u, d).

- Strangeness S:
  The strangeness S is a quantum number carried only by quarks of the *strange* flavor. The convention is simple: for the q_s this quantum number is simply $S = 1$.

---
[6] Werner Heisenberg (5 December 1901 to 1 February 1976) was a German theoretical physicist who made foundational contributions to quantum mechanics and is best known for asserting the uncertainty principle of quantum theory (taken from wikipedia).

1. Introduction

For the anti-quark $\bar{q}_s$ it is $S = -1$. All other quarks have $S = 0$. This quantum number is conserved in strong and in electromagnetic interactions.

With these quantum numbers, the classification of the mesons can be realized as follows: As with the baryons, certain multiplets are defined, ordering the mesons in groups of the same parity and the total angular momentum J^P. The *pseudo-scalar* mesons (Figure 1.3) are the lightest mesons and have:

$$J^P = 0^-$$
$$L = 0 \,,\, S = 0$$

The so-called *vector* mesons are heavier and have the quantum numbers $J^P = 1-$ with $L = 0$ and $S = 1$.

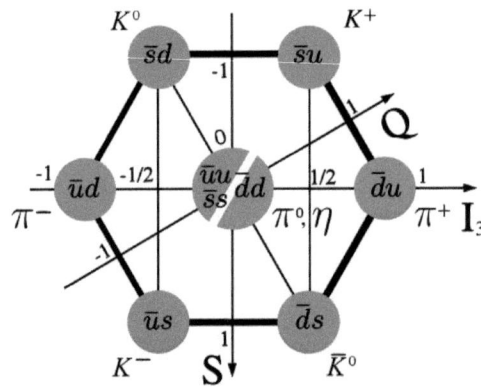

Figure 1.3.: The meson octet for $J^P = 0^-$.

If one takes only the lightest three quarks into consideration the following total number of possible $q\bar{q}$ states is allowed:

$$3 \otimes 3 = 8 \otimes 1$$

Because of the similar masses of u, d and s quarks mesonic particle states will mix. As the s quark is heavier than the u, d quark, the mixing of $s\bar{s}$ with $u\bar{u}$ and $d\bar{d}$ is consequently less pronounced.

The main important properties of the π^0, η and ω meson are listed in Table 1.4 and 1.5, because these mesons were of main interest for this work.

Meson	I^G	J^P	Mass	Life time	relevant decay modes	
π^0	1^-	0^-	134.97 MeV	$8.4 \cdot 10^{-17}$s	$\gamma\gamma$	98.798 %
					$e^+e^-\gamma$	1.198 %

Table 1.4.: Some properties of interest of the π^0-meson [19].

Meson	I^G	J^P	Mass	Life time	relevant decay modes	
η	0	0^-	547.81 MeV	$5.0 \cdot 10^{-19}$s	$\gamma\gamma$	39.38 %
					$3\pi^0$	32.5 %
					$\pi^+\pi^-\pi^0$	22.7 %
					$e^+e^-\gamma$	0.6 %
ω	0^-	1^-	782.59 MeV	$8.3 \cdot 10^{-23}$s	$\pi^+\pi^-\pi^0$	89.1 %
					$\pi^0\gamma$	8.92 %
					$e^+e^-\pi^0$	$7.7 \cdot 10^{-4}$

Table 1.5.: Some properties of the η and the ω-meson [19].

The η-meson

As in this work several decay channels of the η-meson were investigated, with the main focus on the Dalitz decay, some more information about the η shall be given. The η-meson was discovered in pion-nucleon collisions at the Bevatron[7] in 1961. It is characterized by the quantum numbers $I^G(J^{PC}) = 0^+(0^{++})$ and has a life time of $(5.0 \pm 0.3) \cdot 10^{-19}$s. With spin zero and a negative parity it is part of the light pseudo scalar meson octet (Figure 1.3).

The physically observed η-meson is a mixed state of the octet and the singlet states η_8 and η_1: the mixing is described by a mixing angle $\theta \approx -\sqrt{2/3}$ [33]:

$$\eta = \eta_8 \cos\theta - \eta_1 \sin\theta$$

The quark content[8] of the η-meson can be described as follows:

$$\eta_8 = \frac{1}{\sqrt{6}}(u\bar{u} + d\bar{d} - 2s\bar{s})$$
$$\eta_1 = \frac{1}{\sqrt{3}}(u\bar{u} + d\bar{d} + s\bar{s})$$
$$\eta = \frac{1}{\sqrt{6}}(u\bar{u} + d\bar{d} - 2s\bar{s})$$

The strongest[9] allowed decay mode is the second-order electromagnetic transition $\eta \to \gamma\gamma$. The prominent decay channels are listed in Table 1.5. All decay modes via the strong and electromagnetic interactions are forbidden in the lowest order. E.g., the decay into $\pi^0\pi^0$ is forbidden due to the P and CP invariance. Furthermore, the decay into $4\pi^0$ does not occur for the same reason. The electromagnetic decay $\eta \to \pi^0\gamma$ is totally

---

[7] The Bevatron was a historic particle accelerator -specifically, a weak-focusing proton synchrotron- at the Lawrence Berkeley National Laboratory which began operation in 1954.
[8] In respect to the pure $SU(3)$.
[9] The information provided in this paragraph is based on [42].

1. Introduction

suppressed by the conservation of the angular momentum and C invariance. Moreover decays into $\pi^0\pi^0\gamma$ or $\pi^0\pi^0\pi^0\gamma$ are also suppressed, because these violate the C invariance too.

The η-meson is an interesting particle with respect to tests of the QCD symmetries, because the η is an eigenstate of C and CP transformations and due to the blocking of the first order processes, rare decay modes become experimentally accessible. Hence, the investigation of suppressed decay modes, that violate the symmetry, are possible.

1.2.3. The Masses of Hadrons

The mass of a proton is approximately 1 GeV, which is roughly 60 times larger then the combined mass of the three current quarks[10] $\sim$ 15 MeV, that are responsible for the protons quantum numbers and charge. This fact reveals that the origin of hadron masses is not trivial and still needs to be solved. It is common to depict the large additional mass as arising from the kinetic energy of the quarks and gluons and their mutual interaction.

In order to calculate the masses of hadrons correctly, the QCD-Lagrangian needs to be solved for large distances and small energies.

$$L_{QCD} = \bar{\psi}_q(i\gamma^\mu D_\mu - M)\psi_q - \frac{1}{4}G_{\mu\nu}G^{\mu\nu} \qquad (1.12)$$

Unfortunately perturbation theory can not be successfully applied in this energy regime because of the increasing coupling strength. An alternative ansatz is given by the symmetries of the QCD-Lagrangian.
In the following the basic concepts of chiral symmetry will be explained and the conclusions will be discussed.

Chirality is the term for the orientation of the spin of a particle in relation to its momentum vector. Chirality is a conserved quantity only for massless particles. The reason for this is, that for any particle with mass a transformation can be applied, which will affect and invert the momentum vector relative to the particles spin.
In the quantum field theory, chiral symmetry stands for a symmetry of the QCD-Lagrangian under which the left-handed and right-handed parts of Dirac fields transform independently. The transformation of the chiral symmetry can be divided into a component that treats the left-handed and the right-handed parts equally, and a component that actually treats them differently. The former is known as vector symmetry and the latter is called axial symmetry.

Within the chiral limit, and assuming the quarks were massless, the QCD-Lagrangian is invariant under chiral transformation. This assumption is valid because the masses

---
[10]In case of the proton the three current quarks are *uud*.

of the current quarks (u,d) are negligible small compared to the masses of hadrons. If chiral symmetry were also maintained in the hadronic regime, every hadron would possess a chiral partner of the same mass. Chiral partners can only be particles of the same spin but opposite parity. Examples are:

$$\text{Pion: } J^P = 0^- \quad \text{and} \quad \text{Sigma: } J^P = 0^+$$
$$\text{Nucleon: } J^P = \left(\frac{1}{2}\right)^+ \quad \text{and} \quad S_{11}(1535): J^P = \left(\frac{1}{2}\right)^-$$

As the masses of these *partners* very clearly differ[11], it has to be recognized that chiral symmetry is not realized in nature; in other words: it is broken.

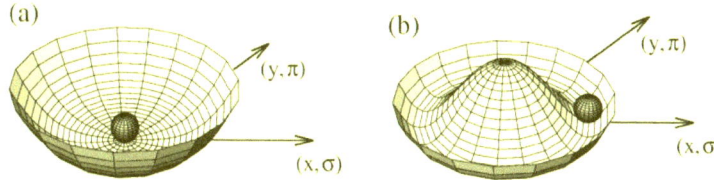

Figure 1.4.: Illustration of the effective potentials in the case of a) no symmetry breaking and b) spontaneous symmetry breaking. The coordinates x and y correspond to the fields $\sigma, \vec{\pi}$ of the strong interaction.

A spontaneous breaking of symmetry is realized, if the symmetry of the QCD-Lagrangian is not equal to that of the ground state. Figure 1.4 shows an illustration of this. In these two pictures the coordinates x and y correspond to the fields σ and $\vec{\pi}$ of the strong interaction. In this case the expectation for the ground state is $\langle \bar{q}q \rangle \neq 0$. For the excitations along the potential valley no energy is needed and thus these states are massless[12]. In contrast, excitations along the radial axis cost energy and therefore belong to states with mass. With this the difference in mass between the nucleon and the S_{11} resonance can be explained; and accordingly in the case of *pion* and *sigma*. What can not be explained by this, is the fact, that the pion and the nucleon have mass (as ground states).

However, the symmetry of the QCD-Lagrangian in nature is only partly realized. Because each current quarks has a mass larger than zero the chiral symmetry is explicitly broken. In principle, this explicit breaking corresponds to a tilt of the potential, which is illustrated in Figure 1.5. Consequently, excitations along the potential valley now cost energy too. As a result of this the pion state as well as the nucleon state become

---

[11] Compare for example the nucleon mass of 1 GeV to the mass of S_{11}, which is 1535 MeV; the mass ratio is larger than 1.5.

[12] These states are the ones for the Goldstone Bosons and (in our example) correspond to the pions.

1. Introduction

massive.

A breaking of symmetry is always connected to a certain order parameter, and in case of the explicitly broken chiral symmetry, this parameter is the so-called chiral condensate $\langle \bar{q}q \rangle \approx (-230\text{MeV})^3$. This condensate decreases with temperature and density as it is shown in Figure 1.6.

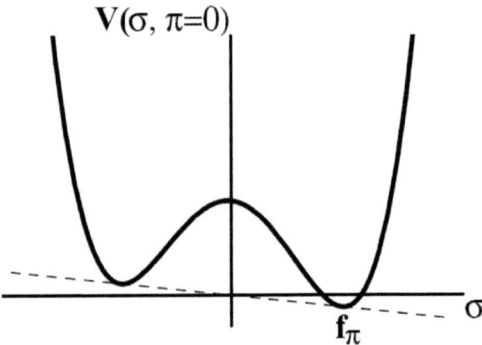

Figure 1.5.: Cut through the three dimensional potential. In case of an explicit breaking of the symmetry, the potential becomes tilted.

When a certain critical temperature or sufficient density is reached a phase transition may take place. By this transition the chiral symmetry will be restored. This could show up in nature as an approximation of the masses of chiral partners (e.g. pion and sigma). Hence, in experiments a broadening or a shift in mass should become observable. In this respect the ω meson is a favored candidate for measurements. The reason for this is the short life time of the ω and the thereby increased probability that the produced ω-mesons will decay within the nuclei.

1.2.4. Electromagnetic form factors

To resolve the inner structure of hadrons, scattering experiments have to be performed. The aim is to measure deviations from the results/predictions of elastic scattering processes on point-like particles. These deviations are expressed as so-called *form factors*.

1.2. Theory

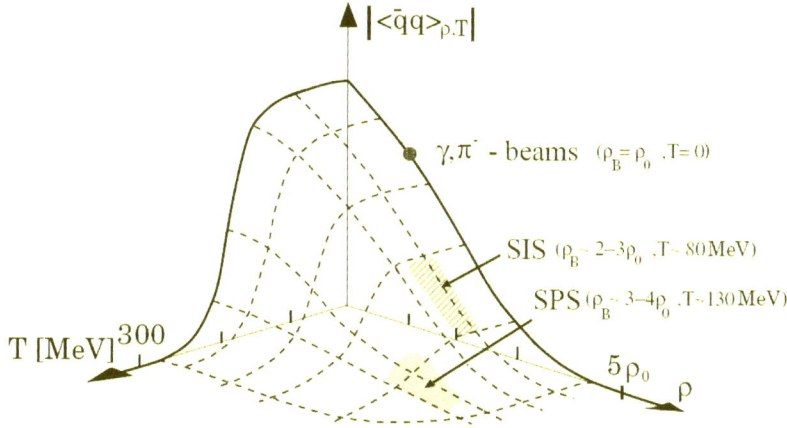

Figure 1.6.: Temperature and density dependence of the chiral condensate corresponding to the Nambu-Jona-Lasinio model [28].

Classical form factors

In order to get a basic understanding we will first focus on the elastic scattering of two charged and point-like particles, which is known as *Rutherford*[13] scattering. While for large impact parameters the structure of the second particle does not play an important role, for small distances a difference in the scattering becomes observable, due to the charge distribution of the second particle (target). This modification is called *form factor*. It depends only on the momentum transfer $\vec{q} = \vec{p} - \vec{p'}$. Hence, the Rutherford cross section formula has to be multiplied by the form factor:

$$\left(\frac{d\sigma}{d\Omega}\right)_{\text{Mott}} = \frac{\alpha^2(\hbar c)^2 \cos^2(\frac{\theta}{2})}{4E^2 \sin^4(\frac{\theta}{2})} \cdot \left(F(q^2)\right)^2 = \left(\frac{d\sigma}{d\Omega}\right)_{\text{point like}} \cdot \left(F(q^2)\right)^2 \quad (1.13)$$

The form factor can be obtained by calculating a Fourier transformation of the charge distribution. Hence, for a charged and point-like particle the form factor is equal to 1.0.

In particular form factors can be determined by measurements of cross sections, which thereafter have to be compared to the theoretical predictions of these cross sections for point like particles. It should be mentioned that not the form factor itself but the square

---
[13]Ernest Rutherford, 1st Baron Rutherford of Nelson, (30 August 1871 - 19 October 1937) was a British chemist and physicist who worked as a pioneer in the field of nuclear physics. In 1908 he was awarded the Nobel prize.

1. Introduction

of the absolute value of the form factor can be determined in this way. A form factor can contain imaginary parts.

Form factors appear not only in scattering processes, but also in electromagnetic production and annihilation processes. In the former case the form factor is space-like, which means that only transfer of momentum will take place ($q^2 < 0$), in the latter case it is time-like as only transfer of energy occurs ($q^2 > 0$). The corresponding Feynman-graphs are shown in Figures 1.7 and 1.8.

Dalitz decays provide a good opportunity to measure electro magnetic transition form factors, which will be discussed in the following.

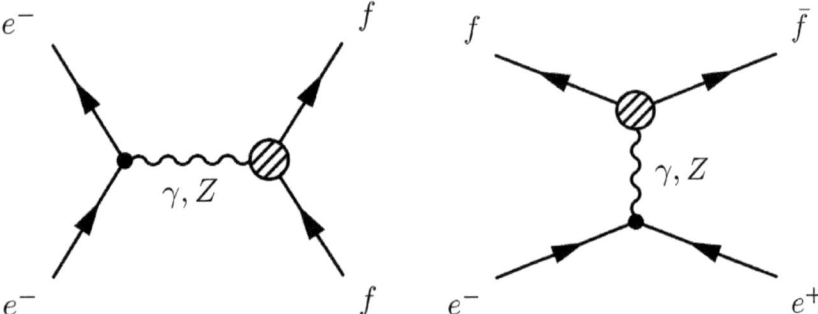

Figure 1.7.: Feynman graph of an electron scattering process. The form factor is measured in the space-like area ($q^2 < 0$).

Figure 1.8.: Feynman graph of a pair production. The form factor is measured in the time-like area ($q^2 > 0$).

Transition form factors

Pseudo scalar neutral mesons of a certain C-parity can couple to another vector meson: $P \to V + \gamma$. The form factor at the vertex is then called *transition form factor* (Figure 1.8). For the η-Dalitz decay $\eta \to \gamma + \gamma^* \to e^+e^-\gamma$ the form factor of the η-γ^*-vertex is meant. This form factor $F(m^2_{e^+e^-}, m^2_\gamma, m^2_\eta)^2$ depends on the invariant masses of the decay products $m^2_{e^+e^-} = q^2$ and $m^2_\gamma = 0$ as well as on the mass of the meson itself m_η. In this case the momentum transfer q^2 corresponds to the invariant mass of the produced lepton pair (e^+e^-).

The distribution of the invariant mass of the two charged leptons (e^+e^-) can be calcu-

1.2. Theory

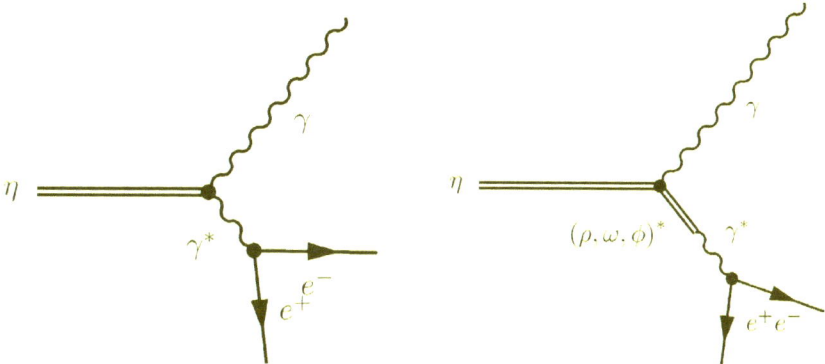

Figure 1.9.: Feynman graph of the η-Dalitz decay. The transition form factor is determined at the $\eta\gamma\gamma^*$-vertex. In this picture the η-Dalitz decay is shown in the picture of the QED.

Figure 1.10.: The η-Dalitz decay in the Vector Meson Dominance model. The virtual γ^* couples to a virtual vector meson.

lated for all Dalitz decays of pseudo scalar mesons (P) in terms of QED [31]:

$$\frac{d\Gamma(P \to l^+l^-\gamma)}{dq^2\Gamma(P \to \gamma\gamma)} = \frac{2\alpha}{3\pi} \cdot \left[1 - \frac{4m_l^2}{q^2}\right]^{\frac{1}{2}} \cdot \left[1 + 2\frac{m_l^2}{q^2}\right] \cdot \frac{1}{q^2} \cdot \left[1 - \frac{q^2}{m_p^2}\right]^3 \cdot |F_p(q^2)|^2$$

$$= \left(\frac{d\Gamma}{dq^2}\right)_{QED} \cdot |F_p(q^2)|^2 \quad (1.14)$$

If only electromagnetic effects among point like particles would play a role, then the form factor would be $|F_p(q^2)|^2 = 1.0$ (Figure 1.9). As also effects of the strong interaction should be present, the form factor should clearly differ from one, $|F_p(q^2)|^2 \neq 1$. Hence, the transition form factor can be determined via an experimental measurement of $\frac{d\Gamma(P \to l^+l^-\gamma)}{dq^2\Gamma(P \to \gamma\gamma)}$ divided by be the QED prediction $\left(\frac{d\Gamma}{dq^2}\right)_{QED}$.

The accessible range of the form factor in case of the *Dalitz* decay is kinetically limited by the mass of the decaying meson, the masses of the produced charged leptons as well as the mass of the third particle m_x. It is $2m_l \leq q^2 \leq m_M - m_x$ (compare to Figure 1.12). In case of the η-Dalitz decay m_x corresponds to m_γ, which is zero. This is different for the Dalitz decay of the ω-meson ($\omega \to e^+e^-\pi^0$); here m_x is equal to $m_\pi^0 \approx 135$ MeV.

Figure 1.13 shows a plot of a calculated transition form factor of the η-Dalitz decay, which has been calculated within a field theoretical treatment. This approach is based on a new, recently introduced counting scheme[14][35] and treats pseudoscalar and vector

---
[14]S. Leupold, stefan.leupold@fysast.uu.se, Uppsala Universitet, Sweden.

1. Introduction

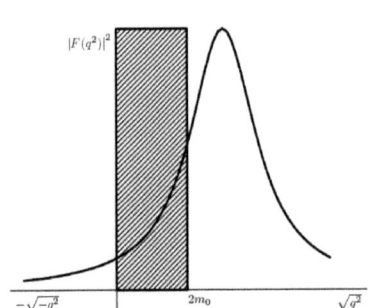 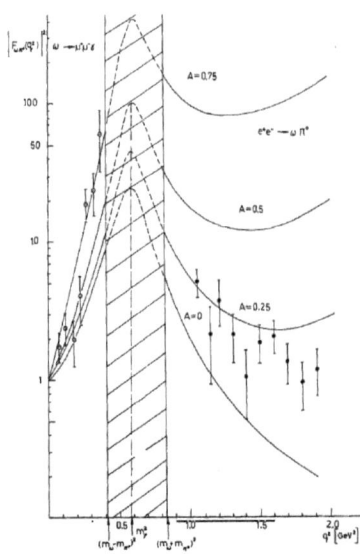

Figure 1.11.: A scheme plotting the characteristics of the form factor for charged pions. The space-like area of negative q^2 can be measured via electron scattering; whereas the time-like area ($q^2 > 2m_\pi$) can be investigated in annihilation experiments. The shaded area can not be investigated in experiments. This Figure and Figure 1.12 are taken from [31].

Figure 1.12.: Measured form factor of the ω-meson in the time-like regime. For $0 < q < m_\omega - m_{\pi^0}$ the form factor can be measured via the analysis of the ω-Dalitz decay; whereas in the area of $q > m_\omega + m_{\pi^0}$ this investigation can be realized by analyzing the production of $\pi^0 + \omega$ in (e^+e^-)-annihilation experiments [31]. Form factors in the area, that is kinetically forbidden (shaded), can be calculated via a dispersion-relation [4].

mesons on the same footing. The calculated data points for this plot were provided by C. Terschluesen[15].

1.2.5. The Vector Meson Dominance Model

In the **V**ector **M**eson **D**ominance model (VMD) the coupling of a photon to a hadron is realized via a coupling to a vector meson as an intermediate state. This concept can

---
[15] C. Terschluesen, Diploma Thesis, University of Giessen, carla.terschluesen@uni-giessen.de

1.2. Theory

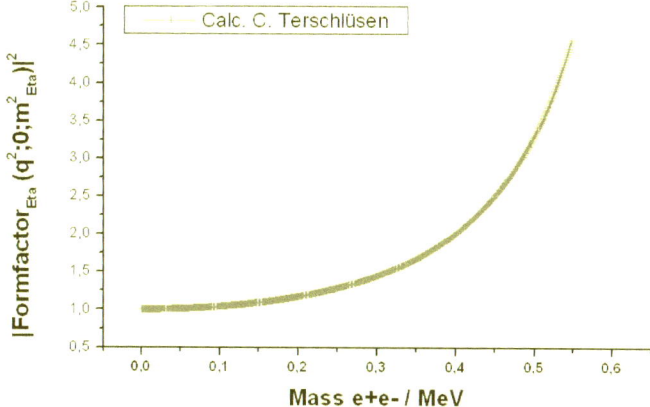

Figure 1.13.: Calculation of the transition form factor of the η-Dalitz decay.

be demonstrated in the parton model, because the photon does interact with one quark of the meson only. Thus a vector meson is created at the $\bar{q}q$-vertex (Figure 1.14).

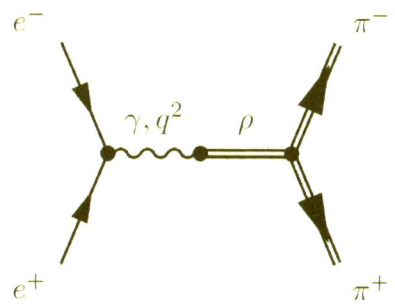 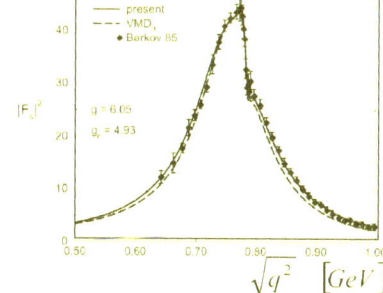

Figure 1.14.: Feynman graph for the investigation of the $\pi^{\pm}$-form factor in the annihilation of e^+e^-.

Figure 1.15.: Measured time-like π form factor in e^+e^- annihilation (Figure 1.14) compared to a VMD prediction [28],[5].

The VMD provides a very accurate prediction for the production of pions in e^+e^- annihilation processes. Figure 1.15 shows a comparison between a measurement and a VMD prediction. One can clearly realize, that the production cross section for pion meson is

1. Introduction

dominated by a resonance at ∼ 770 MeV. This resonance is given by an interference of the ρ and the ω-meson [28],[5].

Concerning the transition form factors the VMD-effects lead to form factors that differ from the value of 1.0 (compare to Figures 1.9 and 1.10). The vector mesons that contribute to the transition form factor are the ρ, ω and the ϕ-meson. The form factors can be parameterized by a one-pol approximation:

$$F(q^2) = \frac{1}{1 - \frac{q^2}{\Lambda^2}} \qquad (1.15)$$

The only parameter in this one-pol approximation is the slope b of the form factor for $q^2 = 0$:

$$b = \frac{dF}{dq^2}\bigg|_{q^2=0} = \Lambda^{-2} \qquad (1.16)$$

Whereas Λ corresponds to the mass of the vector meson.

1.3. Former Experiments and Results

Measuring the electromagnetic transition form factors of light mesons (π^0,η,ω) is a difficult task. Nearly all investigations performed in previous experiments were limited by low statistics; the only exception in this respect is the result of NA60 [12], which has been published recently. In the mentioned NA60 experiment ≈ 9000 decays of $\eta \to \mu^+\mu^-\gamma$ were successfully reconstructed. The result is shown in Figure 1.17.
Another prior experiment, in which the same η-decay was analyzed, is the Lepton-G experiment (≈ 600 counts).

On the one hand analyzing $\eta \to \mu^+\mu^-\gamma$ provides a big advantage as no γ-conversion into e^+e^- contributes to the background. On the other hand the decay into a $\mu^\pm$-pair is further suppressed. Furthermore the measurement is restricted to a certain mass regime, because the mass range below twice the μ-mass is kinematically forbidden.
The results of the Lepton-G experiment (concerning the η-meson) correspond to the VMD prediction. This is different for the ω-meson (60 counts). The latter result shows a discrepancy with the VMD. Thus further experiments are of importance. The Figures 1.18 and 1.19 present the results of the Lepton-G experiment.

The experiment with the (former) highest statistics in the analysis of the $\eta \to e^+e^-\gamma$ Dalitz decay was measured with the SND detector at the VEPP-2M-Collider and was published in 2001 [1].
In this investigation the decays $\phi \to e^+e^-\eta$ and subsequently $\eta \to e^+e^-\gamma$ were analyzed exclusively. The Figure 1.16 shows the result of the determination of the η-Dalitz form

1.3. Former Experiments and Results

factor.

Table 1.4 lists all experiments of interest concerning the investigation of the η-Dalitz decay with a minimum statistic of 50 counts. Prior experiments with even less statistics are discussed in [31]; These early investigations (e.g Jane et al. [24]) had too small statistics and thus the results contained large uncertainties.

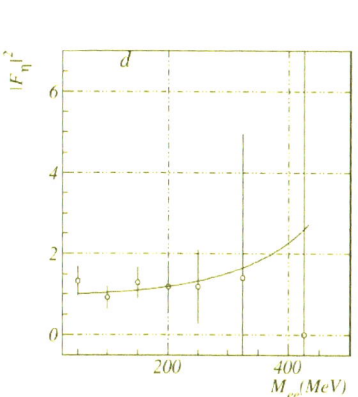

Figure 1.16.: Transition form factor of $\eta \to e^+e^-\gamma$ measured by the SND experiment [1].

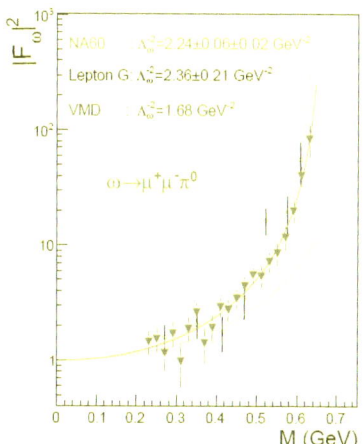

Figure 1.17.: Recent result from the heavy ion experiment NA60 [12].

Experiment	Statistics	Form Factor Slope $[1/GeV^2]$	Pol-Mass $[MeV/c^2]$	Measured
NA 60	9000	$1.95 \pm 0.17 \pm 0.05$	715	$\mu^+\mu^-$
Lepton-G	600	1.9 ± 0.4	720	$\mu^+\mu^-\gamma$
This Work	827	$1.84^{+0.43}_{-0.32}$	740 ± 74	$e^+e^-\gamma$
SND	109	1.6 ± 2.0	790	$e^+e^-\gamma$
HADES	85	$2.2^{+1.2}_{-1.4}$	676	e^+e^-
CB/TAPS@MAMI-B	75	1.99 ± 0.51	708	$e^+e^-\gamma$
VMD	-	1.8	745	e^+e^-

Table 1.6.: Experiments and results of the analysis of $\eta \to e^+e^-\gamma$ and $\eta \to \mu^+\mu^-\gamma$.

1. Introduction

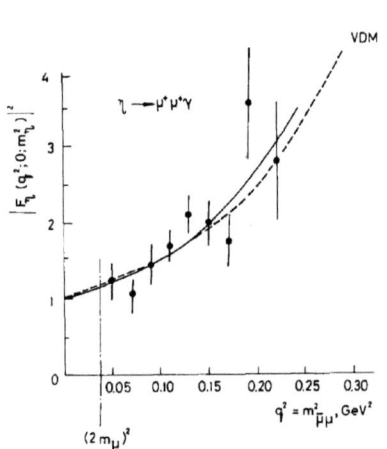 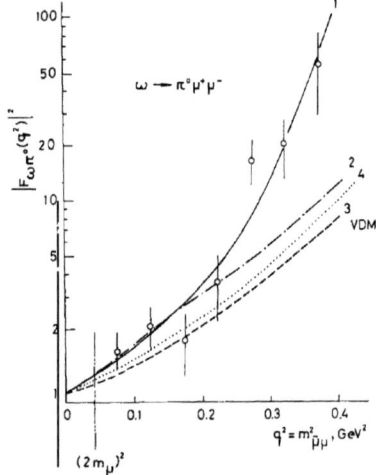

Figure 1.18.: Measurement of the form factor of $\eta \to \mu^+\mu^-\gamma$ by Lepton-G. The solid line is the fit to the data; the dashed curve presents the VMD prediction [31], [8], [13].

Figure 1.19.: measurement of the form factor of $\omega \to \mu^+\mu^-\pi^0$ by Lepton-G. The solid line is the fit to the data; the dashed present the VMD prediction [31], [8], [13].

2. The experimental setup

The experiment was carried out in the A2 experimental hall in the Institut für Kernphysik of the Johannes Gutenberg University, in Mainz, Germany using two beam allocations in June and July 2007. There were several requirements the experimental setup had to fulfill in order to investigate Dalitz decays of neutral mesons. In order to obtain the high accuracy required by our experiment, the high quality electron beam from the MAMI (MAinzer MIcrotron) accelerator in combination with the Glasgow Tagged Photon Spectrometer (Tagger) was used to provide an energy tagged bremsstrahlung photon beam. This photon beam impinged on a liquid Hydrogen target and interacted with the target protons.

In order to cleanly separate events involving Dalitz decays a complete detection and identification of all particles in the final state was necessary. Furthermore a high efficiency and accurate measurements of energy, particle track (angles) as well as timing information were required. To comply with these demands the three component detector system, consisting of Crystal Ball, TAPS and TAGGER was used.

Together the Crystal Ball detector and the TAPS detector cover almost the full solid angle of 4π and thus the detectors provide an acceptance for all η- and ω-momenta. The Particle Identification Detector (PID), situated within the Crystal Ball, was used to register and identify charged particles. For the same purpose the VETO-Wall was used, situated in front of the BaF2-Calorimeter of TAPS in the forward region. In the following sections all experimental components will be described.

2.1. The MAMI accelerator facility

MAMI, the **MA**inzer **MI**crotron [25], [22], is a three-stage racetrack microtron (RTM), beginning with an injector linac supplying electrons with 3.96 MeV total energy and ending with an 855 MeV output beam. A recent upgrade of MAMI (MAMI-C) is the HDSM, the harmonic double-sided microtron. Using the HDSM the maximum beam energy is increased to 1604 MeV [11]. MAMI produces a beam current of up to $100\mu A$. Whereas a synchrotron-accelerator has certain cycles of electron-filling, accelerating and extracting, which always result in a beam of bunched electrons, MAMI provides a continuous electron beam with a duty factor of nearly 100%. It supplies the electron beam for any of the experimental halls (A1, A2, A4, X1) situated at the KPH-Mainz research facility.

2. The experimental setup

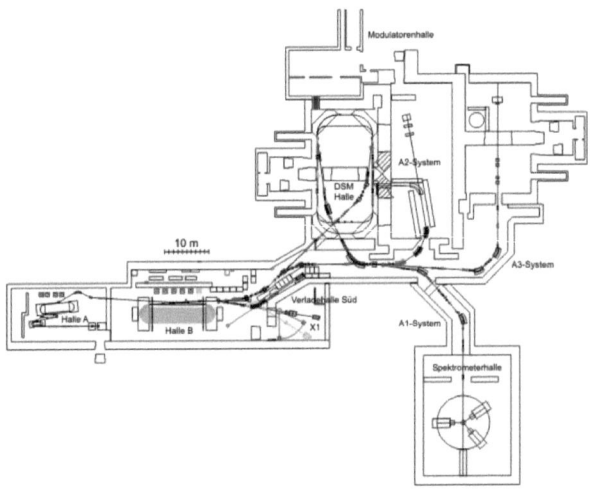

Figure 2.1.: Floor plan of the MAMI-C accelerator facility including the experimental halls A1, A2, A4, X1.

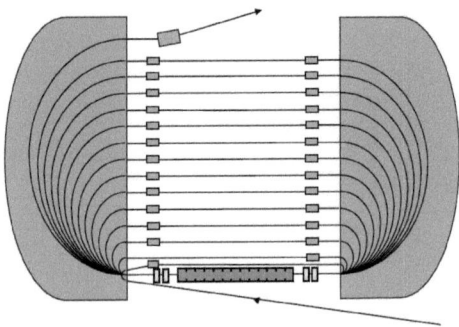

Figure 2.2.: A RTM showing the increased path radius with increasing energy.

Figure 2.1 shows the floor plan in form of a schematic diagram. Table 2.1 lists the main parameters of MAMI.

In a microtron-accelerator the electron beam re-circulates various times through a single linac using constant-field magnets. As the energy of the electron is stepwise increased, the radius of the curvature of the path through the bending magnets increases too (Figure 2.2). As a result of this, each recirculation loop becomes larger. After the electron beam has been accelerated to a certain energy, a small "kicker" magnet ejects it out of the RTM-pathway into the beam handling system. To ensure that the beam bunches see the same phase of the alternating voltage in the accelerating section, the difference in time between each successive recirculation loop has to be an exact integer multiple of the period of the RF supply to this accelerating section.

The microtons in Mainz provide an exceptional phase stability and a very small energy spread of the final beam, which is a result of its inherent phase correction. All particles of higher energy than the designed energy will have a longer path through the bending magnets and thus will arrive later at the acceleration section than the rest of the particle bunch. Hence these particles will be under-accelerated in the next re-circulation. On the opposite particles that are of lower than the designed energy follow a shorter orbit back to the linac section. Thus they arrive early and will be over accelerated. This technique of continual over/under acceleration of particles keeps the spread of energy to a minimum. The **f**ull **w**idth at **h**alf **m**aximum (**FWHM**) of the final energy spread is only 60 keV.

	RTM	RTM	RTM	HDSM
Input energy [MeV]	3.97	14.86	180	855.1
Output energy [MeV]	14.86	180	855.1	1508
No. of recirculations	18	51	90	43
Energy gain/turn [MeV]	0.599	3.24	7.50	13.9 - 16.64
Flux density [T]	0.1026	0.55	1.2842	0.95-1.53
Maximum current [μ A]	100	100	100	100

Table 2.1.: Main parameters of MAMI as taken from [11].

The output energy of a RTM is given by:

$$E_{out} = E_{Inj} + N \cdot \triangle E \tag{2.1}$$

Where E_{Inj} with $\beta \approx 1$ is the linac injected energy. N is the number of cycles of the electron and $\triangle E$ is the additional increase in energy per cycle. The relationship between $\triangle E$ for the electron and the phase of the cavity oscillation is given by:

$$\triangle E = \triangle E_{max} \cdot \cos \phi_s \tag{2.2}$$

2.2. The Glasgow Photon Tagging Spectrometer

The Glasgow Tagger is used to create a continuous and focused photon beam from the electron beam provided by MAMI-C. The electron beam passes through a thin radiator

2. The experimental setup

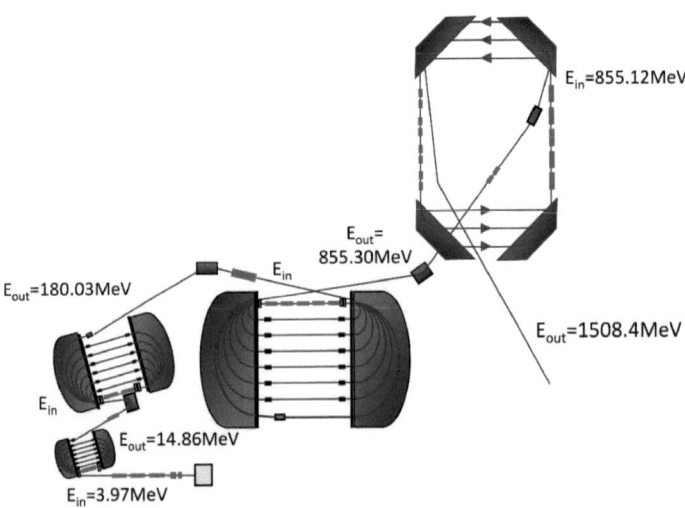

Figure 2.3.: The MAMI-C accelerator. The LINAC, three RTMs and the HDSM are used to accelerate the electron beam up to 1508 MeV.

(copper, diamond), producing photons by the bremsstrahlung process, at energies up to that of the electron beam. These photons keep the original beam direction and impinge on the target. The electrons that have radiated, lost momentum and thus will be bent to larger angles by the magnetic field of the TAGGER than electrons that have not radiated. These impinge on the focal plane detection system of the TAGGER, called the ladder. The electrons that have not radiated are bent into the Faraday cup of the beam dump.

In order to measure the energy of the deflected electrons the position of the electron hit on the focal plane has to be measured (Figure 2.4). Every position on the focal plane corresponds to a certain energy (E_{e^-}). Since the energy of the MAMI electron beam is known (E_0), the energy of the radiated photon can be obtained from the following relation:

$$E_\gamma = E_0 - E_{e^-} \tag{2.3}$$

This equation assumes, that no energy is transfered to the atoms in the radiator [30].

The field strength of the magnet of the Tagger is ≈ 1.8 T. Electrons passing through this magnetic field are bent by the Lorenz force in a circular direction. For a given field strength B and a known curvature R of the particle (example: electron; charge $q = -1$)

2.2. The Glasgow Photon Tagging Spectrometer

the momentum is given by the following equation:

$$p = q \cdot R \cdot B \tag{2.4}$$

The focal plane detector system consists of an array of 353 plastic scintillators. Each of these scintillators is 2 cm wide, 8 cm long and 2mm thick and overlaps with half of its neighbors [21]. As a result of this any tagging electron should trigger two scintillators. Thus the Tagger consists of 353 coincident channels. All events that fired just one channel are rejected in order to reduce the background. The intrinsic resolution of this spectrometer is in the order of 120 keV.

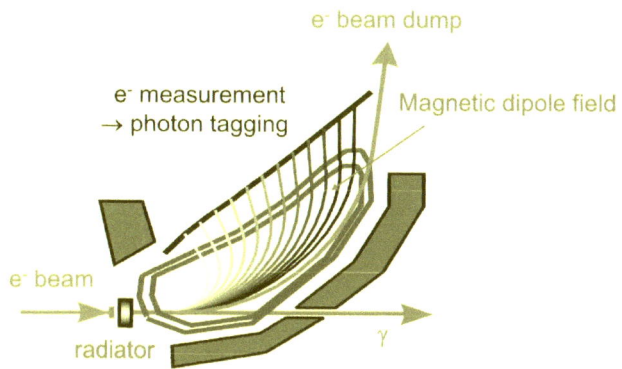

Figure 2.4.: The Glasgow Tagging system.

Later an additional supplementary high resolution focal plane detector (the Tagger-Microscope [43]) was constructed. The purpose of its creation was to exploit the potential of the magnetic spectrometer. It consists of 96 scintillating fibers, and inserted at the true focal plane of the magnet it provides an energy resolution of 400 keV over a limited range of photon energies. However, in the July and June beamtimes in 2007 the Tagger-Microscope was not used.

The Bremsstrahlung distribution follows a $1/E$ shape and due to this, a huge number of low energetic photons are produced. In this respect low energetic photons correspond to high energy tagging electrons. With the intent not to saturate the focal plane detection system, elements at the very high electron energy regime of the focal plane were switched off. This does not bother us since these channels correspond to very low photon energies that are far beyond the production-threshold for η- and ω-mesons.

2. The experimental setup

Every channel of the Tagger is equipped with a small electronics card and comes along with its own photomultiplier tube. The timing resolution of every individual Tagger channel is 0.5 ns FWHM. A resolution of this order is important, because one has to ensure, that the tagged electron energy is attributed to the correct photon, that induced a reaction within the target. Thus a coincidence of the signals from the electron ladder and from the trigger initiated by the Crystal Ball or TAPS detectors is required. This procedure will be discussed in more detail in chapter 4.

2.2.1. Photon Beam and Tagging Efficiency Measurement

Due to the Bremsstrahlung process the photon beam has a $1/E$ distribution. The figure 2.7 shows a typical distribution of the photon beam from the first part of the LH2-run in July 2007. As can be seen, some channels of the Tagger where broken and especially one channel was very noisy. In the analysis this channel was excluded. The photon

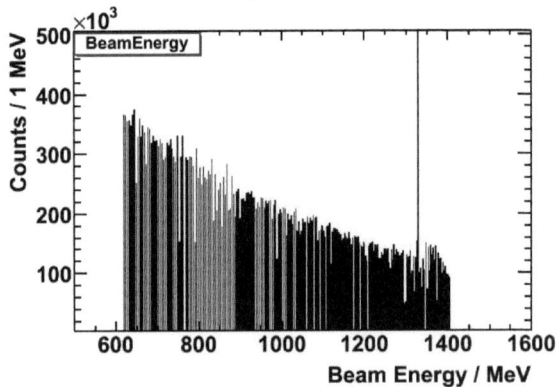

Figure 2.5.: Distribution of photon beam;

beam radius is 2.5 mm. In order to constrain the radius of the beam spot on the target, the beam passes through a lead collimator. During beamtimes a CCD camera is used to observe the shape of the beam spot. The disadvantage is, that the collimation does prevent some 'tagged' photons from reaching the target. In order to perform a cross section measurement, it is important to know the exact number of 'tagged' photons that really impinged on the target; thus a tagging efficiency measurement is necessary. This is done using a special lead glass detector and a very low beam current ($4kHz$). The reason for this is, that this large volume Cerenkov detector has almost 100% efficiency for registering energetic photons at low rates. The detected photons are checked for a coincidence with an electron hit in the tagger. As a result, the ratio (tagging efficiency) can be defined for each tagger-channel:

$$\epsilon_{tagg} = N_\gamma / N_{e^-} \qquad (2.5)$$

where N_γ is the number of detected photons in the lead glass detector and N_{e^-} is the number of electron coincident hits in a certain tagger channel.

The tagging efficiency measurement is carried out daily during a normal beamtime. As the lead glass detector can easily be damaged running at higher rates, it is moved out of the beam during normal data taking. More information about the tagging efficiency can be found in [37].

2.3. The Liquid Hydrogen Target

For this experiment liquid Hydrogen was used as a proton target (Figures 2.6 and 2.7). The target-chassis consists of a cylindrical vessel $48mm$ long and $20mm$ in radius formed from $125\mu m$ Kapton. This vessel is surrounded by several layers of super insulation foil, which are formed from $8\mu m$ Mylar and $2\mu m$ Aluminum to maintain the low temperature (21 K).

The whole assembly is contained in a carbon fiber vacuum tube, which is 1mm thick. The exit window is made of Kapton in order to protect the target cell. Furthermore the Kapton window acts as a scattering chamber to contain the Hydrogen in case of a leakage.

The target gas was liquefied by adiabatic expansion in a special heat exchanger. By this process the Hydrogen was cooled down an condensed. The parameters of the cryotarget system for this operation were 21K and 1080 millibar. The same process can be used in order to produce a liquid Deuterium target. For this slightly different operation parameters (23.5 K ans 1077 millibar) are used. A more detailed description of the target system can be found in [51].

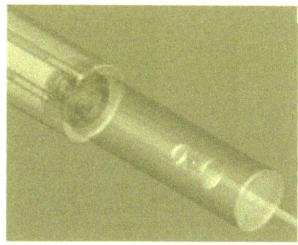

(a) AutoCad representation of the target.

(b) Photo of the Kapton target cell.

Figure 2.6.: Pictures taken from [49].

2. The experimental setup

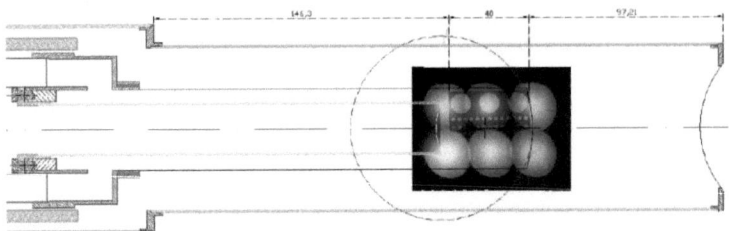

Figure 2.7.: The hydrogen target cell in a technical drawing. Picture was taken from [49].

2.3.1. Probability for a $\gamma \to e^+e^-$-conversion in the target region

A conversion process of a photon into an e^+e^--pair occurs within every medium with a probability given by the radiation length X_0 of the medium. This process contributes to the background in the analysis of final states including an e^+e^--pair. In this work the main focus is on the analysis of the following Dalitz decays $\eta \to e^+e^-\gamma$, $\pi^0 \to e^+e^-\gamma$ and $\omega \to e^+e^-\pi^0$.

As the branching ratios[1] of these decay channels are rather small, conversion processes might generate huge contributions to the background. Thus, it is of importance to estimate the strengh of this effect. If one assumes that a photon, which has been generated in the middle of the lH_2-target, travels orthogonal to the beam axsis and twowards the NaI-Crystals of the Crystal Ball (section 2.4) it has to pass through some material in the target region. This material mainly consists of the 'Liquid Hydrogen Target' and the plastic scintillator of the PID (section 2.4.2). The materials are listed in Tabel A.2 (appendix).

The probability for a photon not to undergo any radiation process is given by:

$$I(x) = I_0 \cdot e^{-\frac{7x}{9X_0}}$$

Thereby I is the intensity, x the thickness and X_0 the radiation length of the material. For a compound of several materials of different thicknesses $I(x)$ has to be calculated as follows:

$$I(x_{total}) = I_0 \cdot e^{-\frac{7x_1}{9X_{01}}} \cdot e^{-\frac{7x_2}{9X_{02}}} \cdot e^{-\frac{7x_3}{9X_{03}}} \ldots$$

With $I_0 = 1.0$ the probability for a conversion of a γ into an e^+e^--pair is given by: $P_{\gamma \to e^+e^-} = 1 - I(x)$. Using the information listed in Table A.2 this probability is:

Without PID: $P_{\gamma \to e^+e^-} \approx 0.6\%$ With PID: $P_{\gamma \to e^+e^-} \approx 1$ for $\theta = 90°\%$

---
[1]$BR_{\eta-Dalitz} \approx 0.6$ %, $BR_{\pi^0-Dalitz} \approx 1.2$ %, and $BR_{\omega-Dalitz} \approx 0.08$ %.

2.3. The Liquid Hydrogen Target

A conversion process in the PID-material does not contribute to the background in the analysis of a final state like $e^+e^-\gamma$ (or $e^+e^-\pi^0$), because the charged leptons must be detected as two separate charged hits, which implies that two PID channels have to be fired and two clusters have to be generated in the calorimeter. Thus, the percentage of photons that generate an e^+e^--pair via the conversion process and which contribute to the background in the Dalitz-analyses is approximately 0.6 %. However, the generated e^+e^--pair needs to be detected. If one assumes a detection efficiency of the CB/TAPS system for photons as 85 %, for $e^\pm$ as 80 % and for protons as 70 %, an η-meson decaying into two photons would contribute via the conversion process to the exclusive analysis of $e^+e^-\gamma$ with a probability of 0.23%. This estimation is close to the result of a Monte Carlo simulation (section 6.4, Figure 6.99).

2.3.2. Number of target protons

The absolute number of target protons is important in order to calculate a cross section. The number of target protons is equal to the number of Hydrogen atoms in the target. Thus this number can be derived from the following equations, knowing the mass number (A) and the molar mass g, the density (ρ) as well as the Loschmidt-number, which is given by $N_L = 6.023 \cdot 10^{23} mol^{-1}$:

$$N_L = A \cdot g \qquad (2.6)$$
$$n = \rho \cdot g \qquad (2.7)$$
$$n = \frac{N_L}{A} \cdot \rho \qquad (2.8)$$
$$\qquad (2.9)$$

Applying this now to Hydrogen:

$$A = 2.02 \frac{g}{mol} \qquad (2.10)$$
$$\rho_{LH_2} = 0.068 \frac{g}{cm^3} \qquad (2.11)$$
$$n_{LH_2} = \frac{6.023 \cdot 10^{23} mol^{-1}}{2.02 \frac{g}{mol}} \cdot 0.068 \frac{g}{cm^3} \qquad (2.12)$$
$$= 2.03 \cdot 10^{22} \frac{H_2 molecules}{cm^3} \qquad (2.13)$$

2.3.3. Other targets

Besides the already mentioned liquid Deuterium using the same DAPHNE cryotarget as the liquid Hydrogen, also liquid Helium3 can be used as target material in the A2-experiments. Furthermore there are several solid target materials available, listed below.

2. The experimental setup

In order to be able to insert solid targets into the beam, the cryostat needs to be moved and replaced by a special system holding the solid target.

Solid target materials

All solid targets for the A2-experiments have a diameter of 30 mm. In the following list all available solid targets are shortly described (these information are taken from [34]).

- Li-Target: the Lithium target has a length of 45 mm and is made of natural Li (92.4 % ^6Li, 7.6 % ^7Li). It had to be produced under an inert protective gas and has to be kept under vacuum during usage, becaus Li strongly reacts with oxygen.

- C-Target: there are three carbon targets available, each of them in form of a disc (length: 10 mm, 15 mm, 25 mm). The carbon target has been used recently (in 2008, by M. Thiel).

- Ca-Target: the calcium target consists of a 10 mm metal disc, which is kept under vacuum to prevent it from oxidizing.

- Pb-Target: the Lead target was a 0.5 mm thin foil consisting of pure ^{208}Pb and was produced by the University of Edinburgh.

- Nb-Target: the available Niob target is of disc-form (0.5 mm).

The system holding the solid targets was developed and produced at the University of Giessen. The main parameters and properties of the targets are listed in table 2.2. In order to normalize a cross section the number of target atoms per cm^2 has to be calculated. This can be done the following way (example Calcium):

$$N_N = \frac{N_L \cdot \rho \cdot L}{A} \qquad (2.14)$$

$$N_N = \frac{6.022 \cdot 10^{23} mol^{-1} \cdot 1.54g \cdot cm^{-3} \cdot 1cm}{40.08g \cdot mol^{-1}} \qquad (2.15)$$

$$N_N = 2.29 \cdot 10^{22} cm^{-2} \qquad (2.16)$$

$$(2.17)$$

The radiation length X_0 of the targets has to be calculated in the following way [19]:

$$X_0 = \frac{716.4g \cdot cm^{-2} \cdot A}{Z(Z+1) \cdot \ln \frac{278}{\sqrt{Z}}} \qquad (2.18)$$

TARGET	^2H	^7Li	^{nat}C	^{nat}Ca	^{208}Pb
Thickness [cm]	4.80	5.40	1.50	1.00	0.05
Density $[g \cdot cm^{-3}]$	0.163	0.534	1.67	1.55	11.35
Surface density $[g \cdot cm^{-2}]$	0.783	2.884	2.51	1.55	0.568
Radiation length $[g \cdot cm^{-2}]$	127.3	82.29	43.26	16.52	6.39
Radiation length [cm]	781.0	154.1	25.9	10.65	0.56
Target thickness [% X_0]	6.14	3.5	5.8	9.4	9.3

Table 2.2.: Properties of the targets materials.

2.4. Crystal Ball System

The Crystal Ball System (**CB**) (Figure 2.8) is composed of the NaI calorimeter, the Particle Iidentification Detector (**PID**) and the Wire Proportional Chamber (**MWPC**). All together these systems cover the same solid angle range viewed from the target at the center of the Crystal Ball. All information of these detector systems combined provide an accurate energy, angle and particle identification in the azimuthal θ-range from $0°$ to $360°$ and in the polar angle (ϕ) from $21°$ to $159°$.

The Crystal Ball consists of two hemispheres, referred to as the 'upper' and the 'lower' hemisphere. Both are made of 25 mm thick Aluminum. The complete assembly has an outer diameter of 66 cm, whereas the inner diameter is only 50.8 cm. The hemispheres are kept under a low pressure in order to prevent damage to the crystals by humidity. To increase the stability a set of steel rope strings between the inner and outer cones is used. Between the upper and lower hemisphere an equatorial plane remains, because of the stainless steel frame and an additional gap of 5 mm. Thus the CB has limited acceptance for particle detection within this area.

2.4.1. The NaI Calorimeter

The NaI calorimeter is a segmented Sodium Iodide detector covering 94% of 4π steradians. This spherical modular detector is based an a fundamental geometric object of a platonic body with 20 identical surfaces (icosahedron). All of the 20 triangles forming that icosahedron have identical properties in length and are referred to as the 'major triangles' of the CB. Each of these triangles is divided into 4 sub triangles (the 'minor triangles'), which consists of 9 Na(Tl)I-crystals of triangular shape. In total the surface of the CB consists of 729 elements. When Crystal Ball was designed and constructed, it was planed to be used in collider experiments. Thus two hexagonal holes exist on opposite sides, once serving as input for the beampipes of a storage ring. Measured in the size of a NaI(Tl)I crystal each hole is of the size of 24 elements; hence the total amount of crystals in the calorimeter is 696, covering a solid angle of 93% of 4π [10].

2. The experimental setup

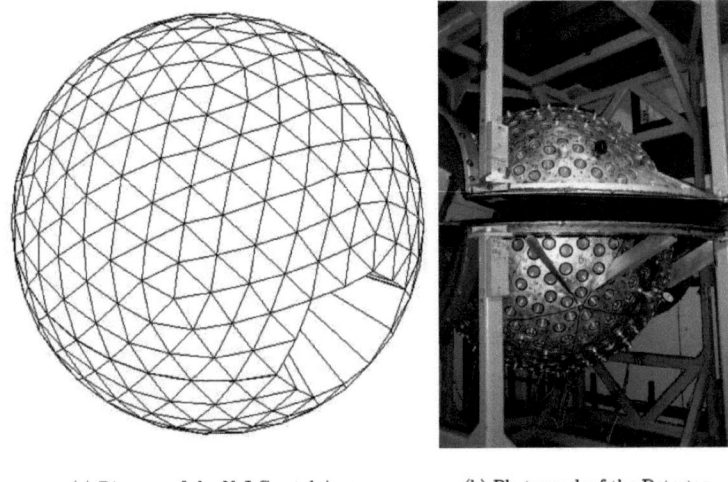

(a) Diagram of the NaI Crystal Array (b) Photograph of the Detector

Figure 2.8.: The Crystal Ball.

Figure 2.9 illustrates the icosahedral setup.

crystal length	15.7 X_0
energy resolution	$\frac{\sigma}{E} \approx \frac{2\%}{\sqrt[4]{(\frac{E_\gamma}{GeV})}}$
time resolution	< 1.5 ns FWHM
polar resolution	$\sigma(\theta) < 3°$
azimuthal resolution	$\sigma(\phi) < \frac{3°}{\sin(\phi)}$
polar acceptance	$20° \leq \theta \leq 160°$
azimuthal acceptance	$1.2° \leq \phi \leq 178.8°$ and $181.2° \leq \phi \leq 358.8°$

Table 2.3.: The properties of the NaI calorimeter.

The shape of crystals is a frustum of a triangular pyramid. Every element is 40.6 cm in length (that corresponds to 15.7 X_0) and has an inner base length of 5.1 cm. The outer base length is 12.7 cm. In order to be able to align all elements in the icosahedrical form as close as possible 11 slightly different crystal shapes had to be used.

The overall stopping power of the Na(Tl)I modules is 425 MeV for protons and 240 MeV for charged pions (π^+, π^-). An early investigation [36] showed, that 98% of the transverse dispersion of a shower will be deposited in an array of 13 crystals.

2.4. Crystal Ball System

Figure 2.9.: Segmentation of the Crystal Ball. The major triangles bordered by a thick black line contain the minor triangles. Each minor triangle corresponds to a Na(Tl)I crystal.

The elements are connected to photomultipliers. For a good optical insulation every crystal is wrapped in 150μm paper and 50μm aluminium foil. All important properties of the NaI calorimeter are listed in table 2.3 and table 2.4.

Density	$3.67 \frac{g}{cm^3}$
Maximal λ of emitted light	410nm
Decay time	230 ns
Light yield	$4 \cdot 10^4$ photons / MeV
Radiation length X_0	2.59 cm
Refraction index n	1.85
Moliere radius	4.3 cm
Minimum ionizing energy	197 MeV
Stopping power protons	425 MeV
Stopping power ch. pions	240 MeV

Table 2.4.: Parameters of the Na(Tl)I crystals.

2.4.2. The Particle Identification Detector

The Particle Identification Detector (PID) is the inner most sub detector in the Crystal Ball. It is approximately 10 cm in diameter and has the form of a barrel (Figure 2.10).

2. The experimental setup

The PID is a $\triangle E/\triangle x$ detector based on plastic scintillators. Each scintillating plastic stripe is 31 cm long, 2 mm thick and 13 mm wide.

The detector was designed and built from 2002-2004. The PID was needed in the CB in order to veto electrons (and other charged particles) and thus to cleanly identify photons. Hence the efficiency of photon identification has been significantly increased since the PID is in use.

Besides the energy information the PID measures the timing of every hit. Since the particle flight path from the target to the Na(Tl)I crystals is very short and the timing resolution of the Na(Tl)I elements is too poor a measurement of the *time-of-flight* is not possible. Thus the identification of particles by the time-of-flight methods is not possible. Hence only the energy information can be used in order to identify charged particles.

Combining the $\triangle E$ information from the PID with the E information measured by the Na(Tl)I calorimeter provides a good separation of protons from charged pions (π^+, π^-), because each type of these particles will be contained in a certain 'band' within the two dimensional histogram plotting dE_{PID} versus E_{NaI}. This technique will be described in more detail in section 4.3.1. As for the analysis of Dalitz decays, a separation of electrons (and positrons) from charged pions is necessary, additional information on the cluster-sizes in the NaI calorimeter as well as timing information (and kinematics in the analysis - momentum balance) have to be exploited.

The energy deposited in the scintillators of the Particle Identification Detector is comparatively small and in the order of 400 keV for a minimum ionizing particle. Still the energy is measurably different for different types of particles of the same total energy.

(a) The Particle Identification Detector, fully assembled in Mainz for testing, before installation.

(b) Final detector position, looking into the CB beam pipe, towards the Photon Tagger.

Figure 2.10.: The Particle Identification detector. Pictures taken from [49].

2.4.3. The Multi-Wire Proportional Chambers

The **M**ulti-**W**ire **P**ropertional **C**hambers (**MWPCs**) were designed to provide a tracking of charged particles such as protons and π^+ or π^-. The detector is placed within the Crystal Ball between the NaI calorimeter and the PID.
The segmentation of the NaI calorimeter is sufficient in order to provide accurate position information for photons, as photons in general fire multiple NaI elements. This allows for a good determination of the exact position of the photon hit (for this the information of the measured distribution of deposited energy is exploited). However, concerning charged particles this method may often not be sufficient, because protons for instance fire only one or two crystals on average. Hence a far superior position information can be obtained using the Multi-Wire Proportional Chambers.

Unfortunaltey the MWPCs could not be used during the beamtimes of interest (July, June 2007), due to a malfunction of the electronic readout of the MWPCs. But still this subdetctor shall be described in more detail.
The MWPCs are filled with a gas mixture of Ar (74.5 %), Ethane (25 %) and Freon (0.5 %). The deposited charge within the chambers is collected from travelling charged particles in a combination of fine wire anodes and thin strip cathodes. These locate the position of the particle's pathway through every individual chamber. The aim of such a device is to obtain the position in at least two wire chambers, because using this, the particle track can be deduced.
The Wire-Chambers are each containd within two coaxial cylindrical Rohacell walls [9]. The walls are of 1 mm thickness and coated in a $25 \mu m$ Kapton film. To gurantee electrical screening a $0.1 \mu m$ thick Aluminum coating is used on the external surfaces of all chamber walls.

The anode wires are made of Tungsten material and are of a $20 \mu m$ diameter. They are positioned at mm intervals around the circumference, parallel to the cylinder axis. The gap between anode to cathode is 4mm. The cathodes are formed by $0.1 \mu m$ thick, 4mm wide stripes made of Aluminum. These are positioned on the internal surface of the Rohacell cylinders. The cathode stripes are wound helically at angles of $\pm 45°$ to the anode wires (Figure 2.11). All of the outer and inner strips cross each other twice along the length of the chamber. The read out is binary based (0 no hit, 1 hit). More properties of the MWCPs are listed in table (2.5).

2.5. The TAPS Detector

TAPS [39] originally had been designed and configured as a '**T**wo/**T**hree **A**rm **P**hoton **S**pectrometer' consisting of 510 Barium Fluoride Crystals equipped with Photmultipliers. However, the crystal split in order to create two new detector setups each serving

2. The experimental setup

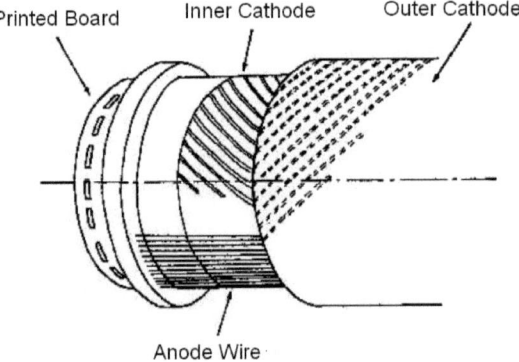

Figure 2.11.: MWPCs: Wire chamber diagram showing cathode winding and anode wires.

azimuthal coverage	$360°$ in ϕ
polar coverage	$21° \leq \theta \leq 159°$
coverage in sterdians	94% of 4π
resolution in z	$200\mu m$
resolution in θ	$1.88°$
resolution in ϕ	$2°$

Table 2.5.: Properties of MWPCs.

as a forward wall in fixed target experiments. Whereas the MiniTAPS situated at CB/ELSA@Bonn consists of only 216 BaF_2 crystals the TAPS detector mounted in the A2-experiment hall in Mainz consists of 384 crystals. The following information will only refer to the TAPS detector of the CB/TAPS@ MAMI experiment in Mainz (Figure 2.12).

BaF_2 was choosen as material because it provides the following properties:

- No hygroscopy and high resistance against radiation damages ([27])

- Very good time response (of the fast component - [27])

- Possibility to perform a pulse shape analysis and to detect neutrons ([40])

2.5. The TAPS Detector

Every of the Barium Fluoride Crystals is capped with a 0.5 mm thick plastic detector, which is used to 'mark' charged particles. All together the plastic detectors form the VETO Wall which will be described in section 2.5.3. TAPS as a forward detector covers the output hole of the Crystal Ball between $0°$ and $21°$ in θ. It is used to detect photons, protons and charged pions in this angle region, which is an important angular range due to the Lorentz boost of particles by the photon beam into forward direction. A upgrade was carried out in two steps in 2008 and 2009. The inner two rings of BaF_2 were replaced by $PbWO_4$. Furthermore the readout of the VETO's was improved and now provides accurate information on the energy ($\triangle E / \triangle x$).

Figure 2.12.: The TAPS Detector [34].

2.5.1. The $BaF2$-Calorimeter

The forward wall calorimeter is made up of 384 BaF_2 detectors [38] and is shown in the diagram in Figure 2.14. The individual elements have a hexagonal cross section, an inner radius of 29.5 mm and are 250 mm in length (Figure 2.13). The individual detectors have an energy resolution of $\frac{\sigma}{E} = 3.7\% E(GeV)^{1/4}$ [38]. As the MWPCs as well as the PID do not subtend the angular region covered by the BaF_2 calorimeter, the granularity and particle identification characteristics of the detector were vital. The BaF_2 is a very sophisticated material providing two different light-output components (a fast and a slow one). Due to it's second component in the light output a **pulse** shape **a**nalysis (PSA) can be performed. This is a suitable method to separate neutrons from photons. In a PSA two integrations of the signals have to be applied; one over the short gate period and one over the long gate period. Plotting the long gate and short gate energies versus each other provides particle 'bands'. One of the bands is attributed to photons, the second band is attributed to neutrons and protons.

Because of the fast rise time of the scintillation pulses, the BaF_2 calorimeter has a timing resolution of approximately $\sigma 200$ ps. As the TAPS-to-target distance is $1,47$ m the **t**ime-**o**f-**f**light (TOF) particle identification method can be used. In order to use

2. The experimental setup

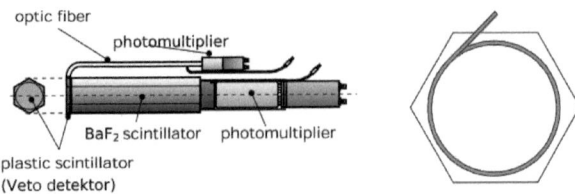

Figure 2.13.: A single TAPS crystal together with a veto detector (right), a light guide and a photomultiplier tube.

TOF in the analysis of A2-data, the time-difference between a signal in TAPS and a signal in the Tagger is plotted versus the energy deposit in TAPS; corresponding to different particle masses distinct bands are seen; more information about this will be given in chapter 4. Table 2.6 and Table 2.7 list the main properties of TAPS and the BaF_2.

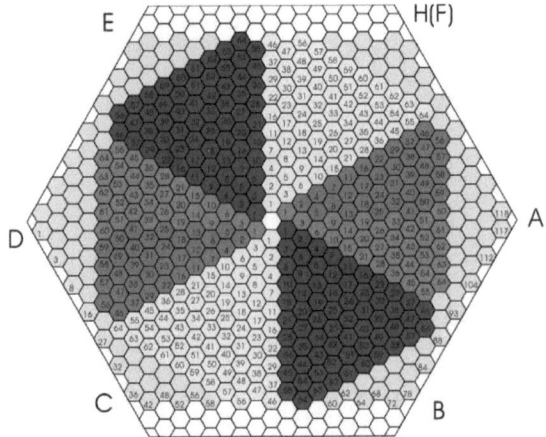

Figure 2.14.: Arrangement of the BaF_2 crystals in TAPS seen in beam direction.

2.5.2. The $PbWO_4$-Upgrade

As this experiment is a fixed target experiment, all in the target produced particles are boosted in forward direction. Hence, the intensity measured in TAPS increases with

2.5. The TAPS Detector

Polar acceptance	$1° \leq \theta \leq 21°$
Azimuthal acceptance	$0° \leq \phi \leq 360°$
Energy resolution	$\frac{\sigma}{E} = \frac{0.79\%}{E(GeV)^{1/2}} + 1.8\%$ [BorisG94]
Stopping power	380 MeV protons | 185 MeV $\pi^{+/-}$ [Boris B89]

Table 2.6.: Main parameters of the TAPS detector.

Density	$4.9 \frac{g}{cm^3}$	Radiation length X_0	2.1 cm
Fast component (FC)	$\tau = 0.8$ ns	Slow component (SC)	$\tau = 620$ ns
Light yield (FC)	$6500 \frac{photons}{MeV}$	Light yield (SC)	$2000 \frac{photons}{MeV}$
Refraction index n	1.56	Moliere radius	3.4 cm

Table 2.7.: Main parameters of the BaF_2 crystals.

lower angles in θ. This means, that the closer a crystal is to the beamline the higher the rate. Figure 2.16 illustrates this situation.
As higher rates increase the statistics without prolonging a beamtime, the limiting factor were the BaF_2 crystals in the inner most ring of TAPS.

In the beginning of 2008 these BaF_2 crystals were replaced by $PbWO_4$, a material standing higher rates and characterized by a higher radiation hardness. In a first upgrade the six inner most BaF_2 elements where removed and replaced by 24 $PbWO_4$ crystals. The new crystals where designed in such a way, that a composition of 4 $PbWO_4$ crystals resulted in the same hexagonal cross section as one BaF_2 element. Thus the new crystals matched the geometrical constraints perfectly and at the same time the granularity could be increased. The drawing in Figure 2.15 illustrates the new design of the inner most ring in TAPS.

As every $PbWO_4$ element is read out by its own photomultiplier the rates per crystal could be decreased by a factor of 4 compared to the situation before. Furthermore the stopping power of $PbWO_4$ is larger than that of BaF_2, due to the heavier elements and their higher atomic numbers (Pb: 82, W: 74). This leads to a shorter radiation length X_0 as well as to a smaller Moliere radius. All important information on the $PbWO_4$ crystals are listed in table 2.8.

End of 2008 another upgrade of the TAPS calorimeter was accomplished, replacing the second ring of BaF_2 against the $PbWO_4$ crystals.

2.5.3. The VETO Wall

In order to distinguish between charged and neutral hits, every element of the TAPS calorimeter has a 5mm thick hexagonal plastic scintillator in front, which is read out

2. The experimental setup

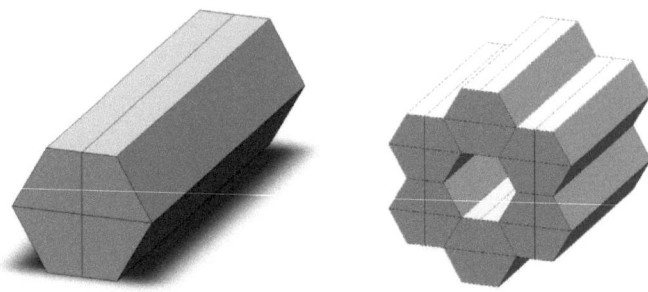

Figure 2.15.: Left: 4 $PbWO_4$ crystals as composition. Right: the new inner Rinf of TAPS.

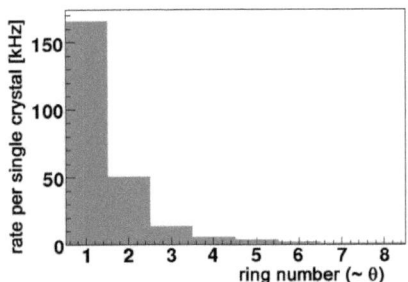

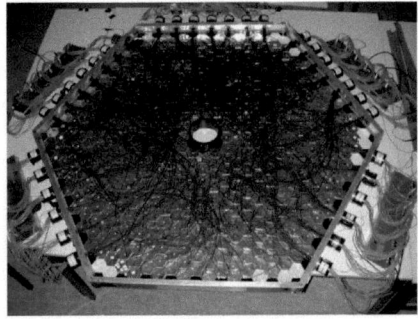

Figure 2.16.: Rates in TAPS. The closer a crystal is to the beamline, the higher the rate.

Figure 2.17.: Photograph of the aluminum frame with all veto detectors and the light guides visible.

with an optical fiber. Each plastic scintillator is made of EJ20n [16] and is only sensitive to charged particles (Type NE 102A).

The 384 Veto modules are mounted in a box consisting of an aluminum frame with a 1 mm thick PVC front plate. They were glued on a 3 mm thick rear plate. To suppress and minimize light loss, each veto was wrapped into a thin aluminum foil; to get all modules light-tight some extra layers of black tape were glued onto them. Figure 2.17 shows a photograph of the complete VETO wall.

The veto modules were read out using a wave length shifting fiber light-guide (Bichron

	$PbWO_4$	BaF_2
Density [g/cm^3]	8.3	4.9
Maximal λ [nm] (FC)	420	220
Maximal λ [nm] (SC)	560	310
Decay time [ns] (FC)	10	0.7
Decay time [ns] (SC)	50	620
Radiation length X_0 [cm]	0.89	2.05
Refractive index n	2.20	1.56
Moliere radius [cm]	2.0	4.3

Table 2.8.: Main parameters of $PbWO_4$ in comparison to BaF_2.

BCF-92). The light-guides were linked to 16-fold photomultipliers (Hamamatsu H6568). In total 33 PM's were installed on the outer side of the aluminum frame [Jan00]. The output of each PM is connected to the input of a 16 fold CFD (Ganelec FCC8). The digitized information, whether a channel fired or not, was given to a coincidence register (LeCroy 4448). A CAMAC backplane was used by the A2-Controller to read out the BitPattern coming from the coincidence registers.

Meanwhile the electronics and read out systems of the VETO Wall were improved. Now it is possible to distinguish between different types of charged particles via the $\triangle E/E$ method, as with the PID. This work highly profited from this improvement, as it made the separation e^+, e^- from π^+, π^- easier and increased the efficiency of proton identification.

2.6. Electronics, data acquisition and read-out

2.6.1. Crystal Ball

Figure 2.18 shows the readout electronics for the Crystal Ball detector in a schematic form. All signals coming from the CB are transfered to active splits. Each split sums up the signal over 16 channels in total, providing the energy sum as global trigger. One of the 16 channel outputs is directly connected to the discriminators; another is delayed by 300 ns and connected to the ADC's. Each discriminator module consists of two discriminators and has individual thresholds for each channel. One discriminator provides the cluster multiplicity information, the other one provides the information for the CATCH-TDC's. The TDC's as well as the ADC's scan the signal continuously; no gate signal has to be provided. In case of a positive feedback from the trigger, which is distributed to all modules, the data buffer of those modules are passed to the next acquisition phase. Processing computers read out these buffers via the VME bus. As the ADC's are VME cards they do not communicate via VME bus nor S-Link. In

2. The experimental setup

order to control and readout the ADC's a special multiplexer (GeSiCA or iMUX) is used.

The amplification of the signal is done using SRC L50B01 photomultipliers in combination with a voltage divider. These voltage dividers are operating with +1500 V anode voltage. To ensure that the energy information for an identical deposition of energy in all crystals is the same, the gain has to be adjusted (using a potentiometer).

An important issue is furthermore, to eliminate long dead times during data acquisition. As the readout-servers are operating with a real time operating system (OS - LINUX), enabling parallel running of software, the data can be read out in a pseudo parallel mode and can be stored to disk. The residual dead time is given by the time between the generation of a trigger signal and its registration by the readout program as well as by the time for the sequential readout of the modules.

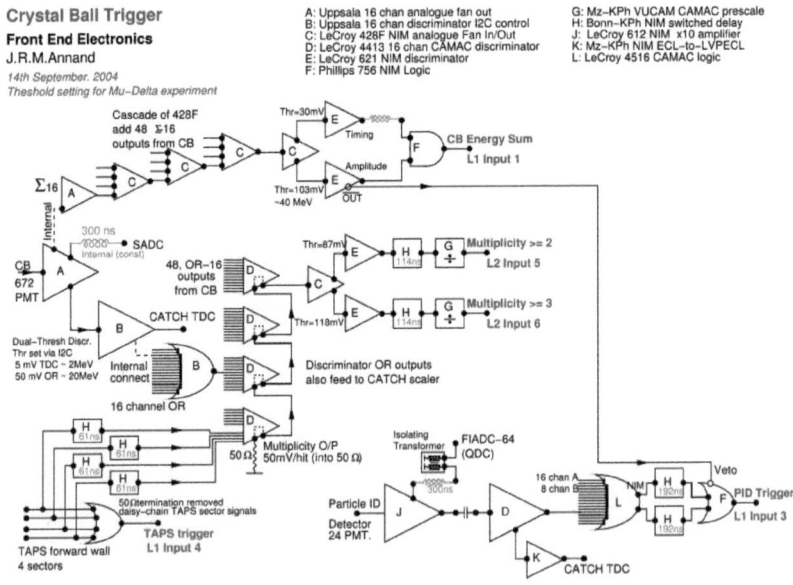

Figure 2.18.: Scheme of the CB readout system.

2.6.2. TAPS electronics

The electronics of TAPS is based on the VME bus-standard. For the readout a CAEN-V876 motherboard is used, combined with a special circuit board mounted on top of the motherboard. All readout-circuits are placed on the so called 'piggyback' (constant fraction discriminators, leading edge discriminators, charge to analog converters, ADCs). All Piggy-boards have a VME-interface and are read out by a VME-CPU via the VME backplane. All electronics, especially the Piggy-Boards, are described in detail in [15].

Crosstalk

The crosstalk describes the influence of neighboring channels on each other on the VME boards. A modified version of the slowcontrol server was used to perform a measurement of cosmics and all channels on the boards were switched on/off in all possible combinations. Comparing the peak position of the TDC the crosstalk between the channels could be determined [Lugert]. Figure 2.19 shows the result of a crosstalk measurement for all BaF_2 elements. This measurement was carried out while TAPS consisted of $510 BaF_2$ crystals.

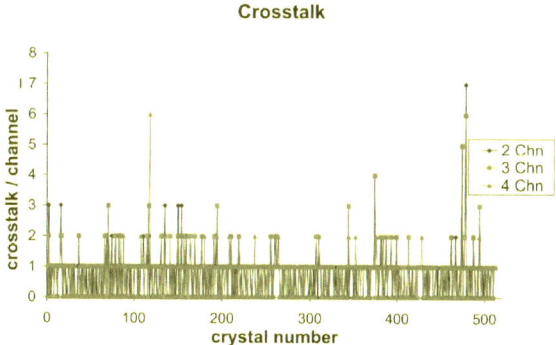

Figure 2.19.: Crosstalk measurement; the BaF_2-crystals (x-axis) are plotted against the maximum crosstalk in the TDC-channels (y-axis).

Computer systems used for data taking

For this experiment VME CPU's (VMIC VMIVME7750) are used. The reasons for this are:

2. The experimental setup

- Elements on the architecture can be changed

- No hard disks are contained; thus in case of a power crash the experimental setup is not affected

- Increase of the number of CPU's is easy

- All CPU's share an identical OS

The VME CPU's are located directly at TAPS in the experimental hall. They are booted via BOOTP and TFTP. The file system (NFS) is mounted from a Linux server placed in a rack in the A2-counting room (server-1). Server-1 runs the BOOTP and TFTP services too. The boot procedure is described in [34]. A plan of the network setup in Mainz referring to the TAPS experiment is shown in Figure 2.20. The server-1 is named **TAPS01**. A second server called **TAPS02** is used to run the MySQL-database and to store cosmic-data.

These two servers are still operating, but they are used as a backup today. Because of their age a new powerful server called **TAPS00** has been installed, which takes over the old servers tasks (2010). Before this another computer (**TAPS03**) was installed too and was used for an online-analysis in 2008 during th ω-runs on Niob and Carbon.

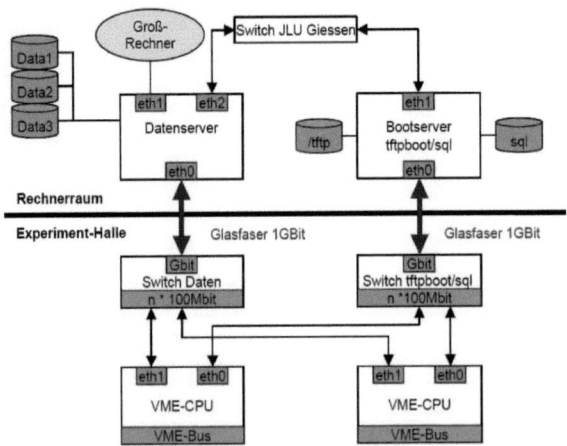

Figure 2.20.: The TAPS computer system

2.7. The Trigger

In experiments like CB/TAPS@MAMI it is not possible to record all registered events, due to the high rates. Moreover, as the reaction between beam and target is a statistical process depending on cross sections of all possible reactions, it is not possible to select just the one particular channel of interest. Hence, a trigger system is needed in order to discriminate between non interesting and interesting events; only the latter ones are saved to disk. Besides sufficiently high rates, the trigger system should provide a good signal/background ratio.

The trigger system used with CB/TAPS@MAMI is divided into two sections, referred to as 1st level (L1) and 2nd level (L2) trigger. L1 is an energy sum trigger (only Crystal Ball) and L2 is a hit-multiplicity trigger. How the energy sum is built, is described in [50]. For the beamtimes of interest (LH_2 runs in June, July 2007) the minimum energy sum was set to 300 MeV.

The hit-multiplicity trigger helps to separate events of interest from non interesting events. Since in this work the Dalitz decays of the η and the π^0 meson were investigated, a multiplicity of $M3+$ was set. Thus every event consisting of at least three hits in the detectors TAPS and/or CB and fulfilling the energy sum requirement in the CB was saved to disk. The η, ω and π^0 Dalitz decays are given by:

$$\eta \to e^+e^-\gamma \qquad (2.19)$$
$$\gamma p \to \eta p \to e^+e^-\gamma p \ (Reaction) \qquad (2.20)$$
$$\omega \to e^+e^-\pi^0 \qquad (2.21)$$
$$\gamma p \to \omega p \to e^+e^-\pi^0 p \to e^+e^-\gamma\gamma p \ (Reaction) \qquad (2.22)$$
$$\pi^0 \to e^+e^-\gamma \qquad (2.23)$$
$$\gamma p \to \pi^0 p \to e^+e^-\gamma p \ (Reaction) \qquad (2.24)$$

The analysis was based on an exclusive analysis, demanding the detection of all four particles in the final state (including the backscattered proton). Thus a L2 trigger of $M4+$ could have been chosen. As the Dalitz analysis was planed to be cross checked by the analysis of $\eta \to \gamma\gamma$ (including the backscattered proton), the trigger had to be set to $M3+$. Furthermore the decays $\pi^0 \to \gamma\gamma$ as well as $\eta \to \gamma\gamma$ are used for calibrating the energy of the calorimeters, since the invariant masses of π^0 and η are well known. In order to have sufficient $\gamma\gamma$-events for calibration purposes, an additional multiplicity trigger 'M2' scaled down by a factor F ($F \in 2, 4, 6, 8$) was used.

A scheme of the trigger logic is shown in Figure 2.21. The CB-area, which is covered by 16 channels of a certain discriminator, defines the cluster size. The 'OR' signals of the 16 channels of the discriminator are translated into NIM signals. All these signals are collected by four LeCroy (4413) discriminators 'B'. The shown module 'H' delays

2. The experimental setup

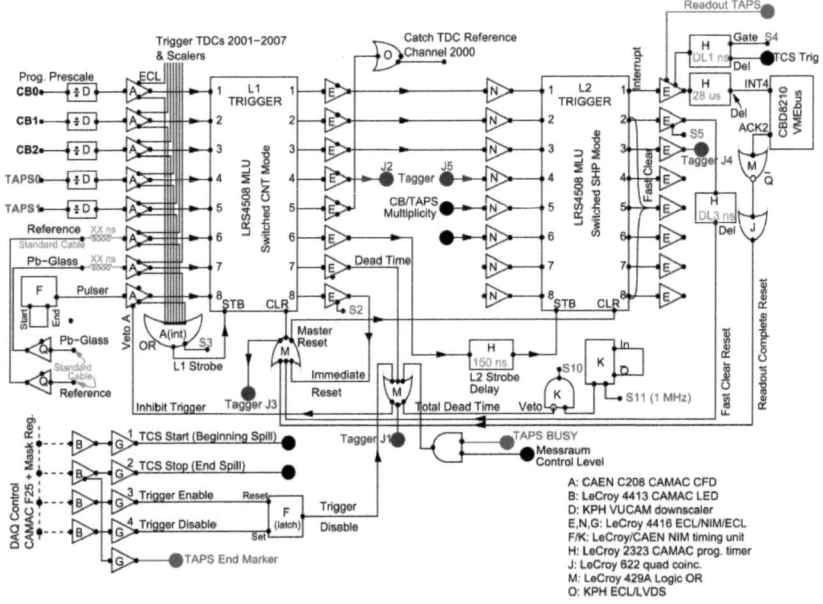

Figure 2.21.: Trigger logic of the combined CB/TAPS system.

the signal for the L2-trigger. The module 'D' can be used to scale down the number of accepted triggers by software. The eight TAPS-segments are combined to four logic segments which contribute to the multiplicity.

The event latch in Figure 2.21 has the task to recognize the L1-pattern, to generate signals for the readout and to detect a confirmation of the L2-trigger. If a event is 'valid', the readout is initialized. For more information about the trigger system consult [14] and [34].

2.8. Beam-time overview

As the focus in the research of the Giessen A2-group is set on the investigation of Dalitz decays of neutral mesons as well as on the study of in-medium effects of the ω-meson,

2.8. Beam-time overview

special experiments with different targets were performed at the research facility KPH[2] in Mainz, Germany. The electron accelerator MAMI-C provided an electron beam with up to 1.5 GeV, which was used to generate a γ-beam via the bremsstrahlung process. The Crystal Ball, TAPS, and Tagger detector systems were used in a combined mode to collect data.

For the analysis of the Dalitz decays three beamtimes, using a LH_2 target, were performed and analyzed; in 2007 two run periods were realized using the maximum γ-beam energy of 1408 MeV on an unpolarized proton target. Unfortunately the scalers can not be read out from the data of the first run-period; thus the corresponding data files could not be used for the determination of cross sections. For analyses in which high statistics were important (e.g. determination of form factors) both beamtimes were used. As the third beamtime was performed in order to produce and investigate π^0-mesons at the threshold, a lower γ-beam energy of 885 MeV was used.

Concerning the studies of in-medium effects three run-periods were accomplished using the targets Nb, C, Nb and the highest available γ-beam energy at that time (1408 MeV).

All analyzed beamtimes were calibrated by the A2-group of Giessen. The beamtimes are summarized in Table 1.7.

Beamtime	Target	Energy [MeV]	Hours
2007_06	LH_2	1408	197
2007_07	LH_2	1408	160
2008_04	Nb	1408	450
2008_06	C	1408	225
2008_08	Nb	1408	329
2008_12	LH_2	885	100

Table 2.9.: Overview on (analyzed) beamtimes.

---

[2]Institut für Kernphysik at the Johannes Gutenberg Universität in Mainz, www.kph.uni-mainz.de .

3. Calibration

An accurate calibration is an important requirement for subsequent analyses. As in Giessen the Dalitz Decays of neutral mesons as well as in-medium effects of the omega meson [47], [32] were to be investigated, a strong focus was set on performing a very precise calibration of all components (NaI-calorimeter, BaF_2-calorimeter, Veto-Wall, PID, TAGGER).

To be able and observe any in-medium effects of the ω-meson, it has to be ensured, that observed effects do not result from an insufficient calibration. Further on very precise cuts on the kinematics have to be applied - especially in the analysis of the the η-Dalitz decay. Thus an enormous effort was put into the development of flexible routines for the planed calibrations. In a first step a special calibration-class (TA2Calibration) for the AcquRoot[1] (software for data analysis) was developed in C++. The purpose of this class was to preprocess the calibration in selecting certain events, applying loose cuts and writing preselected data to files. In a second step these files were analyzed using special calibration-macros (C++/ROOT based). Most calibration-macros first had to be developed. Some macros, based on older versions from 2006 could be modified.

All calibrations that were performed in Giessen (since 2007) are listed in table 3.1. The complete calibration procedures for each detector component will be described in this chapter.

3.1. NaI Energy Calibration

The calibration of the Na(Tl)I calorimeter was done in three steps. The initial alignment of the hardware gain of the Crystal Ball was performed using a radioactive source; this procedure is referred to as 'basic' calibration and is described in section 3.1.1. In a second step the experimental data was analyzed exploiting the information from $\pi^0 \to \gamma\gamma$ events. As the mass of the π^0 meson is well known, its decay into two photons provides a very good possibility to calibrate the detectors in energy. Unfortunately it was found, that the relation of channel-to-energy did not follow just a linear function. Thus in a third step a quadratic energy calibration had to be performed. For this, the η-decay into $\gamma\gamma$

---

[1] AcquRoot: the main software for analyzing A2-data. This program will be described in more detail in chapter 4.

3. Calibration

	2007-06	2007-07	2008-4	2008-06	2008-08	2008-11	2008-12
Target	LH_2	LH_2	Nb	C	Nb	^3He	LH_2
NaI-Energy	√	√	√	√	√	√	√
NaI-Time	√	√	√	√	√	√	√
NaI-TimeWalk	√	√	√	√	√	√	√
PID-Energy	-	-	-	-	-	-	-
PID-ϕ-Corr.	√	√	√	√	√	√	√
BaF2-Energy	√	√	√	√	√	√	√
BaF2-Time	√	√	√	√	√	√	√
VETO-Energy	√	√	√	√	√	√	√
VETO-Time	√	√	√	√	√	√	√
TAGGER-Energy	√	√	√	√	√	-	-
TAGGER-Time	√	√	√	√	√	-	-

Table 3.1.: Overview over all calibrations performed in Giessen (since 2007); these are available for download [7]. The calibrations marked with a green √ were done by the author himself.

was analyzed. As with the π^0-meson the mass of the η-meson is well known. Hence, the investigation of $\eta \to \gamma\gamma$ provided all needed information for the second order correction in energy, which will be described in section 3.1.2.

3.1.1. Basic Calibration

In order to perform an accurate calibration of the NaI channels, the gain for each channel has to be set correctly. In a first and basic step this was done using a radioactive source $^{241}Am/^9Be$, which was installed in the middle of the Crystal Ball (target location). The source emitted beyond others a very well defined γ-ray of $\approx$ 5 MeV. The ADC spectra of all Na(Tl)I channels were measured. The peak position of the detected photoline (Figure 3.1) was adjusted via a potentiometer in such a way, that the center position of the peak became the same for all Na(Tl)I channels. With this, a basic calibration was performed.

The radioactive source contained ^{241}Am and ^9Be; the latter likes to absorb α particles. ^{241}Am has an half-life of 432.2 years and emits α particles of two different energies (12 % 5.443 MeV and 85 % 5.486 MeV). During this radioactive decay the daughter nucleus ^{237}Np is created:

$$^{241}_{95}Am \to\, ^{237}_{93}Np + \alpha \qquad (3.1)$$

3.1. NaI Energy Calibration

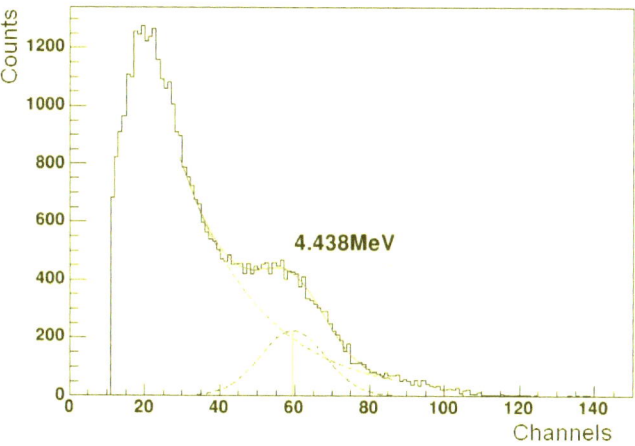

Figure 3.1.: ADC spectrum after using a $^{241}Am/^9Be$ radioactive source for energy (gain) calibration [50].

The Beryllium absorbs α particles of this energy and forms a compound[2] nucleus $^{13}C^*$, which decays into three channels:

$$^9Be + \alpha \rightarrow\, ^{13}C^* \rightarrow \begin{cases} ^{12}C^* & +n \\ 3\alpha & +n \\ ^8Be & +\alpha + n \end{cases} \qquad (3.2)$$

The $^{12}C^*$ is dominant in production and at the same time of importance for the basic calibration. Because it is an excited state, it will decay into its ground state ^{12}C emitting a photon (γ) with an energy of 4.438 MeV.

$$^{12}C^* \rightarrow\, ^{12}C + \gamma \qquad (3.3)$$

These photons are used for the basic calibration.

3.1.2. Linear Calibration

During the experiment the raw energy information obtained from the ADC channels of the NaI crystals was recorded to disk in form of a data stream. In principle the ADC channel represents the electron yield from the photomultiplier after a particle hit the

---
[2] The reaction is $^{B}e(\alpha,n)^{12}C^*$.

3. Calibration

crystal and deposited an amount of energy larger than the threshold; this threshold is given by a certain ADC setting. The electron yield is directly correlated to the amount of deposited energy [27].

In order to perform a proper calibration, a transformation of the channel information into an energy value is needed. In general, the correlation between energy and read out ADC channel number is given by linear function:

$$E = gain \cdot channel + offset \tag{3.4}$$

The linear calibration procedure is based on one precondition. This is, that the pedestal, which is the channel that corresponds to zero energy (in MeV), was set properly. If this is the case, the offset is known. Thus only one point is needed to obtain the gain factor, which can be derived from the measured pion mass ($\pi^0 \approx 135$ MeV). As the invariant mass is not energy, but energy should be calibrated, this procedure demands some preparatory work.

- Because the π^0 decays into $\gamma\gamma$ ($BR \approx 99\%$), events containing two neutral hits have to be selected and investigated.

- Assuming E_1 is the energy of the first photon and E_2 the energy deposited in a number of NaI modules by the second photon, the invariant mass can be calculated (with α as the measured opening angle between the two neutral hits):

$$m^2 = 2 \cdot E_1 \cdot E_2 \cdot (1 - \cos\alpha) \tag{3.5}$$

- In the next step the calculated invariant masses of every event are plotted against the channel numbers of the central crystal of the NaI-clusters, which were fired by the two photons.

- After that a macro[3] is used to create projections onto the mass-axis for every NaI-channel. In the following step the π^0-mass peak in the projected histograms (for each NaI-channel) has to be fitted. Figure 3.4 shows the fits of 50 NaI-channels (of one iteration).

- From this a new gain-value can be obtained via the following equation:

$$gain_{NEW} = gain_{OLD} \cdot \frac{m_{\pi^0}^{PDG}}{m_{\pi^0}^{measured}} \tag{3.6}$$

$$m_{\pi^0}^{PDG} = 134,976 \frac{MeV}{c^2} \tag{3.7}$$

This procedure is an iterative one. The gain setting will converge to its proper value with every iteration. Figure 3.2 illustrates the reason why several iterations were needed.

---
[3]CalibNaILinear.C, based on a macro written by D. Werthmüller.

3.1. NaI Energy Calibration

On the left is shown, the invariant mass of the π^0 per detector channel number after the first iteration; on the right (Figure 3.3) the same after 27 iterations.

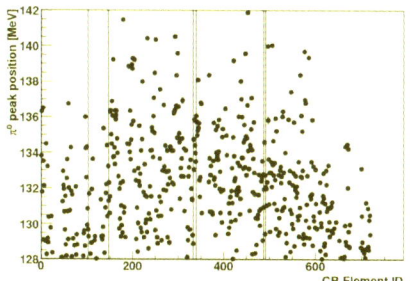 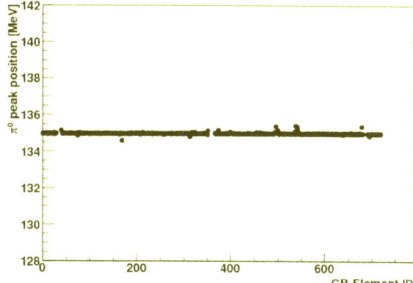

Figure 3.2.: Reconstructed π^0 mass per channel of the *NaI* after the first iteration.

Figure 3.3.: Reconstructed π^0 mass per channel of the *NaI* after the 27th iteration.

It was found, that an accurate calibration of the energy can not be done separately from a time calibration: as a stepwise improved time calibration allows to apply more and more stricter cuts on the prompt peak in the time spectra, the signal-to-background ratio in the π^0 analysis is increased and thus the π^0 signal can be fitted with more and more accuracy. The time calibration of the *NaI* will be described in section 3.2. Hence, during every iteration a time and an energy calibration-step had to be accomplished.

One main reason for the demand for 'iterations' is the following: the gain factors for the energy are obtained channel by channel. But for every channel, the calibration point (m_{π^0}) is reconstructed from two neutral hits. Each hit fires a whole cluster of crystals, and the summed energy information is used for the calibration procedure of just one channel (which is always the index-crystal of the cluster - the crystal that detected the highest amount of energy); The neighbors are ignored.

Beamtime	06-Part 1	06-Part 2	06-Part 3	06-Part 4
Diff. [%]	1.53	1.67	1.55	1.97
Beamtime	07-Part 1	07-Part 2	07-Part 3	07-Part 4
Diff. [%]	1.35	1.53	1.74	1.51

Table 3.2.: Difference of the reconstructed η-mass to the PDG value (for LH_2-Beamtimes 06/2007 and 07/2007)

3. Calibration

Furthermore it is a fact, that **two** photons (that means two clusters of crystals) form a pion, but only one is used in equation 3.4. The energy information of the other cluster is assumed as correct, which is not exactly the case. Thus many iterations are needed in order to converge to an accurate gain value for each channel.

The data of each beamtime were split into four parts of similar size. Although the position of the reconstructed π^0 mass peak in the data was located at an accurate value after the linear calibration, the invariant masses for η-meson (and the ω-meson) were off. In addition it was found, that the peak position varied from part to part of the data. Table 3.2 lists the deviation of the reconstructed η mass from the PDG mass ($mass_\eta^{PDG}$ = 547.85 MeV). As for the ω meson the difference in the invariant mass were even worse ($\approx 3.2\%$ on average), a non-linear calibration was needed.

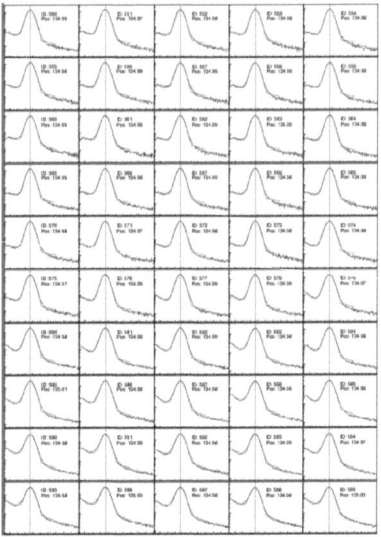

Figure 3.4.: The fits to the π^0 mass distributions of 50 NaI channels.

3.1.3. Second order Calibration

The assumed linearity (equation 3.4) of the transformation of channels into energy is only valid in the range of low invariant masses (π^0 mass). It has been found that using only the linear calibration the η-mass and the ω-mass both differ from their PDG values. As in this work η-decays as well as ω-decays were to be investigated a more accurate calibration method was needed. Thus a second order correction had to be applied:

$$E = c_1 \cdot channel + c_2 \cdot channel^2 \quad (3.8)$$

For this a second calibration point is necessary, which is the mass of the η. The η meson does also decay to $\gamma\gamma$ and its mass is well known. Furthermore the huge number of produced η mesons ensures that there is enough statistics available in the data in order to perform a proper calibration. The questions is know, how can the correction factors c_1 and c_2 be determined. Using the equation 3.5, the real pion mass is given by:

$$\begin{aligned}
m_{\pi^0}^2 &= 2 \cdot E_{\gamma 1} \cdot E_{\gamma 2} \cdot (1 - \cos\alpha) \\
&= 2 \cdot (c_1 \cdot E_{\gamma 1meas} + c_2 \cdot E_{\gamma 1meas}^2) \cdot (c_1 \cdot E_{\gamma 2meas} + c_2 \cdot E_{\gamma 2meas}^2) \cdot (1 - \cos\alpha) \\
&= 2 \cdot E_{\gamma 1meas} \cdot E_{\gamma 2meas} \cdot (1 - \cos\alpha) \cdot (c_1 + c_2 \cdot E_{\gamma 1meas}) \cdot (c_1 + c_2 \cdot E_{\gamma 2meas}) \\
&= m_{\pi^0 meas}^2 \cdot (c_1 + c_2 \cdot E_{\gamma 1meas}) \cdot (c_1 + c_2 \cdot E_{\gamma 2meas})
\end{aligned}$$

3.1. NaI Energy Calibration

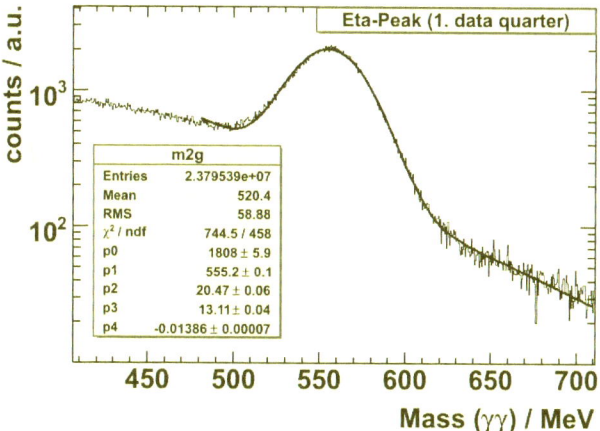

Figure 3.5.: Result of the linear calibration procedure. The η-mass is still off.

The correction factors c_1 and c_2 can not be set event wise. Hence, the values for the factors have to be determined by the photon energies E_1 and E_2. Using an averaged energy $\langle E_{\gamma\pi^0}\rangle$ leads to:

$$m_{\pi^0}^2 = m_{\pi^0 meas}^2 \cdot (c_1 + c_2 \cdot \langle E_{\gamma\pi^0}\rangle)^2 \qquad (3.9)$$

The question is how to determine $\langle E_{\gamma\pi^0}\rangle$? Two simple options are available, the geometric mean and the arithmetic mean.

1. geometric mean: $\qquad \langle E_{\gamma\pi^0}\rangle = \sqrt{E_{\gamma 1} \cdot E_{\gamma 2}} \qquad (3.10)$

2. arithmetic mean: $\qquad \langle E_{\gamma\pi^0}\rangle = \dfrac{E_{\gamma 1} + E_{\gamma 2}}{2} \qquad (3.11)$

The geometric mean was chosen for all non-linear NaI-calibrations performed in Giessen (Table 3.1). The π^0 in the index of E in the equations above should indicate, that this photon was used to reconstruct a pion. Now the same procedure is applied on the η-meson:

$$m_\eta^2 = m_{\eta meas}^2 \cdot (c_1 + c_2 \cdot \langle E_{\gamma\eta^0}\rangle)^2 \qquad (3.12)$$

To obtain the correction factors the equations 3.9 and 3.12 need to be solved. In order to do so, the ratio of the PDG-mass value of the pion to its measured value is calculated

3. Calibration

(the same for the eta meson):

$$R_{\pi^0} = \frac{m_{\pi^0 PDG}}{m_{\pi^0 meas}} \quad (3.13)$$

$$R_{\eta^0} = \frac{m_{\eta^0 PDG}}{m_{\eta^0 meas}} \quad (3.14)$$

With this the equations can be solved and the factors can be obtained:

$$c_1 = R_{\pi^0} - c_2 \cdot \langle E_{\pi^0 \gamma} \rangle \quad (3.15)$$

$$c_2 = \frac{R_{\pi^0} - R_{\eta^0}}{\langle E_{\pi^0 \gamma} \rangle - \langle E_{\eta \gamma} \rangle} \quad (3.16)$$

This procedure was developed by K. Makonyi[4] and modified by B. Lemmer [32]. During the development of this calibration technique two very important questions came up, which shall be discussed and answered:

- Can one directly obtain the gain factors for the ADC channels instead of those for the energies ?
- How to calculate a correction factor in case the previous run was already corrected ? Can this method be used iteratively ?

The second question can be answered with 'yes', since this procedure was applied iteratively. Concerning the first question, the answer is simple. Assume k_1 and k_2 were the gain factors for the ADC channels:

$$E = c_1 \cdot E_{meas} + c_2 \cdot (E_{meas})^2 \quad (3.17)$$
$$= c_1 \cdot (k_1^{old} \cdot ch + k_2^{old} \cdot ch^2) + c_2 \cdot (k_1^{old} \cdot ch + k_2^{old} \cdot ch^2)^2 \quad (3.18)$$
$$= [ch] \cdot (k_1^{old} \cdot c_1) + [ch^2] \cdot (c_1 \cdot k_2^{old} + c_2 \cdot (k_1^{old})^2) + \cdots \quad (3.19)$$

With new values for k_i:

$$k_1^{new} = k_1^{old} \cdot c_1 \quad (3.20)$$
$$k_2^{new} = c_1 \cdot k_2^{old} + c_2 \cdot (k_1^{old})^2 \quad (3.21)$$

These new values for k_1 and k_2 can be saved to the configuration file of the crystals of the NaI-calorimeter.

After 9 iteration steps this non-linear calibration of the NaI has been accomplished with a very accurate result. For all parts of the beamtimes of interest the deviation in the reconstructed η mass from the PDG value was 0.6% on average. For the ω-meson a slightly higher average deviation of 0.75% has been achieved. The following pictures (Figures 3.6, 3.7, 3.8 and 3.9) give an overview over these results.

---
[4]Karoly Markonyi, karoly.markonyi@exp2.physik.uni-giessen.de

3.2. NaI Time Calibration

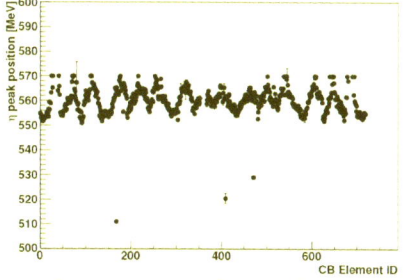

Figure 3.6.: Reconstructed η mass per channel after the first iteration.

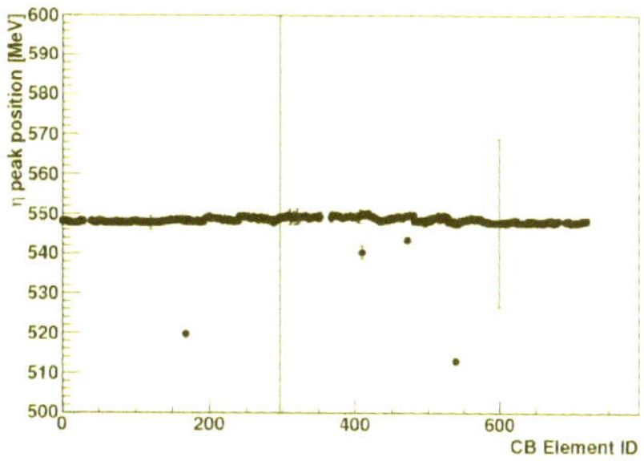

Figure 3.7.: Reconstructed η mass per channel after the 9th iteration.

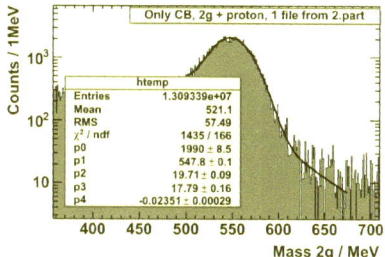
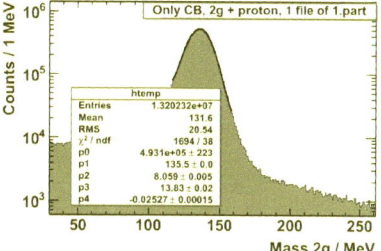

Figure 3.8.: Reconstructed η mass after calibration.

Figure 3.9.: Reconstructed π^0 mass after calibration.

3.2. NaI Time Calibration

During data taking the timing of every signal from each individual NaI-crystal was recorded by a CATCH TDC and saved to disk. Due to the nature of the CATCH TDCs, a certain and fixed value for the channel-to-time conversion had to be set. This setting was equal for all crystals and had the value 117.71 ps / channel. However, to use the timing information of the NaI-calorimeter for an analysis, it was necessary to align all the prompt peaks of all NaI-elements.

For this purpose again $\pi^0 \to \gamma\gamma$ events were analyzed. Similar to the the energy calibration, the prompt peak in the timing spectrum of each NaI-element was fitted by a Gaussian. Thus the mean values were determined. In the next step a constant offset was applied to each channel in order to shift the timing peaks to an equal and common point in the timing spectra. This was done using the macro 'CalibCBTime.C', which

3. Calibration

was developed by the author. Of course the resolution of the timing becomes better the more iterations of the procedure were applied. This kind of calibration should not be done without a proper energy calibration, as the latter allows to cut on the invariant mass of the π^0 and thus helps to remove background. Hence, the signal-to-background ratio can be improved and the fitting routine will work more precisely. In total 7 iterations of this procedure were applied.

Unfortunately, the time calibration of the Crystal Ball is not that simple. The already accomplished calibration via an alignment of all channels by shifting the offset can only be regarded as a first step. Still the time difference between two photons (or any other two particles) hitting the Crystal Ball was broadened by an effect called *time walk* (Section 3.2.1).

3.2.1. CB Time Walk Correction

The reason for the existence of *the time walk* effect is the design of the Crystal Ball. As the TDCs that are used with the CB are CTACH TDCs, they need a global reference time signal to start. This signal is given by the trigger. As a fact the time of the event, which causes the triggering, depends on the energy of the particle that is triggered.

Every discriminator requires a threshold of deposited energy; but the 'moment' when this threshold is reached depends on the particle energy. Figure 3.10 illustrates this problem.

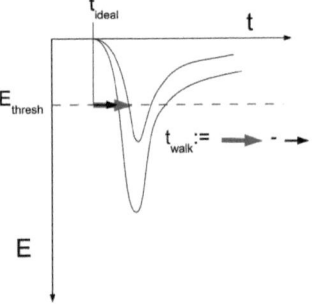

Figure 3.10.: The *time walk*. For a lower signal the trigger threshold is reached later in time.

For low energetic hits this threshold simply needs a longer time to be reached as for high energetic hits. Thus a delay of the trigger signal is produced. The resulting effect on the data by the *time walk* is shown in Figure 3.11. In this 2D histogram a strong curvature of the time versus energy relation can be seen; this is the *time walk* spectrum of a single channel. On the right side (Figure 3.12) the corresponding spectrum for all channels is shown; this 2D histogram was created after the *time walk* correction had been accomplished.

In order to correct for the *time walk*, an energy dependent *time walk correction function* is needed. This function can be obtain via the following procedure:

3.2. NaI Time Calibration

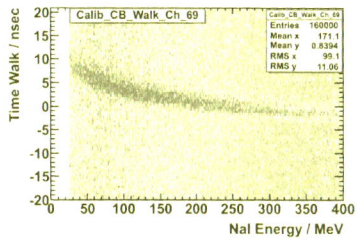 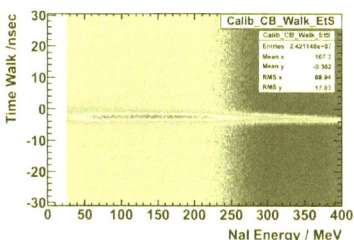

Figure 3.11.: *Time Walk* of NaI-Channel 69.

Figure 3.12.: Time Walk of all channels plotted versus the NaI-Energy after correction.

- A two dimensional histogram has to be generated for each channel, plotting the *time walk* against the channels energy in MeV (e.g Figure 3.11).

- For every energy bin a projection onto the *time walk* axis has to be created. For this a lot of statistics is necessary. After that, the peak in the projected histogram has to be fitted (by a Gaussian function). As a result *one point* (center position) is retrieved.

- All these points have to be fitted by a so called *time walk function* $f_{walk}(E)$, which is the correction function. After that, the retrieved correction parameters of the *time walk function* are written to the NaI-configuration file. This implies, that the *time walk function* is implemented into the decoding routines of the data-analyzer-software. This can simply be done in the following way:

$$Time_{channel_i} = Time_{\text{(offset corrected)}} - f_{walk}(E) \qquad (3.22)$$
$$(3.23)$$

The question is now, which function should be used as $f_{walk}(E)$. Beyond others two functions were investigated:

$$1: \quad f_{walk} = p_0 + p_1 \cdot \left(\sqrt{\frac{p_2}{E}}\right) \qquad (3.24)$$

$$2: \quad f_{walk} = p_0 \cdot e^{p_1 \cdot \sqrt{E} + p_2 \cdot x} + p_3 \qquad (3.25)$$

The first function is a three-parameter function, which is used in the standard Acqu-Root[5] code. In this function p_0 is a shift constant, p_1 corresponds to the 'RaiseTime', p_2 is the threshold and 'E' the energy. This function was used first. Unfortunately the result was not convincing (at least for the liquid Hydrogen beamtimes). Thus the second

---
[5] AcquRoot is the main analysis program, which will be described in detail in chapter 4.

3. Calibration

time walk function was tested, which led to the desired results (Figure 3.12).

After all individual NaI channels were corrected (*time walk*), a re-alignment had to be done to correct the offsets once again, which were affected by the time-walk-correction too. Thereafter all channels were time calibrated. Figure 3.13 shows the spectrum of the time difference of the two photons detected in the Crystal Ball (after a cut on the reconstructed m_{π^0}). The FWHM of the fit to the peak is 3.04 ns, which is a sufficiently good time resolution (for the CB) and comparable to the values found in [32], [29].

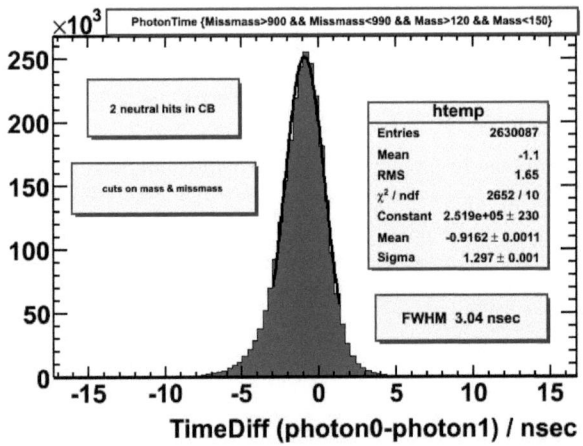

Figure 3.13.: Time resolution of two neutral hits in the CB. The FWHM of the fit to the peak is 3.04 ns.

3.3. BaF_2 Energy Calibration

As with the NaI-crystals, the energy calibration of the BaF_2-calorimeter is a three step process. The crystals pedestal (zero energy) channels were determined by observing the strong peak in the lowest region of the ADC spectra. The channel with the highest value before the first decrease was determined to be the pedestal value. To obtain this zero energy deposit pedestal value, the TAPS pedestal pulser forced a readout of all TAPS ADCs once per second. The result of this was a very narrow peak. Once this was established, the first step of the energy calibration, the cosmic calibration, could be performed.

After the cosmic calibration had been accomplished, a linear energy calibration based on the analysis of the decay $\pi^0 \to \gamma\gamma$ was performed (as with the NaI). The result was an accurate calibration in the mass range of the π^0. Unfortunately it was found again, that the mass of the η meson was off. Thus a second order correction had to be applied according to the procedure with the NaI crystals.

3.3.1. Cosmic Calibration

For this first calibration step the energy deposit of cosmic Myons was exploited, in order to acquire a first (linear) relation between energy and channels. As this relation is assumed to be linear, two points are needed. The pedestal (zero energy) was used as the first point. As second point the mean of the Gaussian distribution of the energy deposit of cosmic Myons was used. The energy deposit of these Myons is the same in all BaF_2 crystals with a peak at 37.7 MeV.

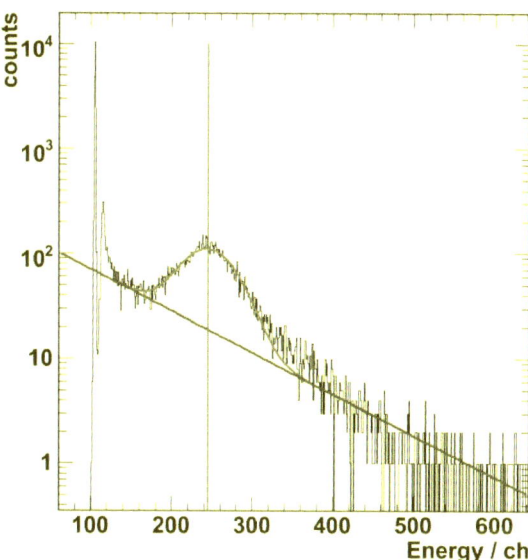

Figure 3.14.: Distribution of energy deposit by cosmic Myons on the passage through a BaF_2 crystal.

Cosmic Myons are minimum ionizing particles, that means their energy deposition does

3. Calibration

not depend on their momentum, but depends only on the absorbing material (BaF_2). The energy loss of a Myon in a BaF_2 crystal of TAPS is approximately 6.5 MeV per 1 cm passage. The mean path length in a BaF_2 crystal of TAPS is 5.9 cm. Thus the average deposit of energy amounts to $\approx$ 38 MeV. More information on this can be found in [45].

The cosmic data used for the calibration were recorded before and after each beamtime. During this, TAPS was used in the stand-alone data acquisition.

3.3.2. Linear energy Calibration

The procedure for the linear calibration using $\pi^0 \to \gamma\gamma$ works in the same manner as for the $Na(Tl)I$ crystals. As TAPS does not cover a large fraction of the solid angle, the statistics of $\pi^0 \to \gamma\gamma$ events with both photons detected in TAPS was very low. Furthermore it is not possible to detect $\eta \to \gamma\gamma$ event only with TAPS. This is because of the large opening angle between the two decay photons in these η-decays. Thus events with one photon in TAPS and one photon in the Crystal Ball had to be used; and hence this required an already accomplished energy calibration of the CB. The rest is exactly the same as with the NaI crystals.

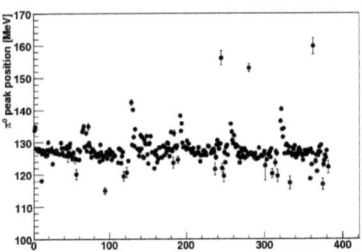

Figure 3.15.: Invariant mass positions of $m_{\pi^0 meas}$ per channel after the fourth iteration.

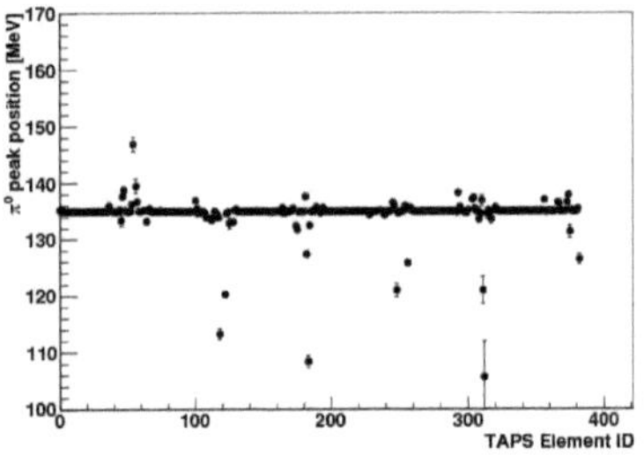

Figure 3.16.: Invariant mass positions of $m_{\pi^0 meas}$ per channel after the final 18th iteration.

For this calibration a program based on a macro written by D. Werthmüllers was used. Some modifications had to be applied. Again this procedure depended on fitting the measured m_{π^0} signal, calculating new gain-factors and writing them to the configuration file. After 18 iterations the result shown in Figure 3.16 was achieved. Some channels were still off. Most of those belonged to the inner most ring of BaF_2 crystals. As these are very close to the beamline, there is always a very strong electro magnetic background; thus the fitting often went wrong because of the bad signal-to-background ratio. Finally the gain factors and offset values of these channels were set manually. For those the average values of all other channels were used.

3.3.3. Second order Calibration

Similar to the NaI-case the linear calibration was not sufficient for the BaF_2 detectors. Indeed the reconstructed π^0 mass was very close to the PDG value, but this did not hold for the η-mass, which was about 10 MeV off on average. This fact is illustrated by Figure 3.17. Hence, as with the NaI detectors a second order correction was needed.

For the TAPS experiment at CB/ELSA in Bonn there was already a solution on how to apply such a correction. Instead of a correction of the energies according to equation 3.8, the energies here were corrected not for individual crystals but as a function of θ and ϕ. Unfortunately this method could not be just copied because of a lot of constraints like the differences in software, setup and reconstruction method of particles. Finally an independ, but slightly similar method was developed.

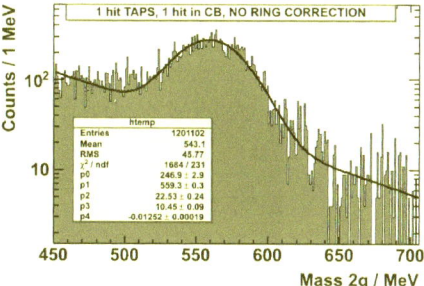

Figure 3.17.: After the linear calibration, the η-mass is still off.

The method described in 3.1.2 (NaI case) could not be used directly, because:

- the statistics of $\eta \to \gamma\gamma$ events in TAPS was far too low (because of the opening-angles-restrictions)

- hence, the statistics per crystal was too low as well

- consequently: events with one hit in the TAPS and one hit in the CB were used and thus new formulas were required.

As B. Lemmer worked out in his Thesis [32], the simplest idea is the following:

$$
\begin{align}
m_{\pi^0}^2 &= 2 \cdot E_{\gamma TAPS} \cdot E_{\gamma CB} \cdot (1 - \cos\alpha) \tag{3.26}\\
&= 2 \cdot (c_1 \cdot E_{\gamma TAPSmeas} + c_2 \cdot E_{\gamma TAPSmeas}^2) \cdot E_{\gamma CB} \cdot (1 - \cos\alpha) \tag{3.27}\\
&= 2 \cdot E_{\gamma TAPSmeas} \cdot E_{\gamma CBmeas} \cdot (1 - \cos\alpha) \cdot (c_1 + c_2 \cdot E_{\gamma TAPSmeas}) \tag{3.28}\\
&= m_{\pi^0 meas}^2 \cdot (c_1 + c_2 \cdot E_{\gamma meas}) \tag{3.29}
\end{align}
$$

3. Calibration

The problem is now again to find an averaged $E_{\gamma\pi^0 meas} =: \langle E_{\gamma\pi^0 meas} \rangle$. A solution is to average over all photons in TAPS, that belong to a two-γ event and fulfill a cut on the π^0-mass. The same is done in order to determine $\langle E_{\gamma\eta meas} \rangle$, only this time a cut on the η-mass has to be applied.

With this equation 3.30 can be written in the following way:

$$m_{\pi^0}^2 = m_{\pi^0 meas}^2 \cdot (c_1 + c_2 \cdot \langle E_{\gamma\pi^0 meas} \rangle) \qquad (3.30)$$

And for the η-meson:

$$m_{\eta}^2 = m_{\eta meas}^2 \cdot (c_1 + c_2 \cdot \langle E_{\gamma\eta meas} \rangle) \qquad (3.31)$$

After that the ratio of the PDG mass values to the measured masses is calculated: $R_{\pi^0} = \frac{m_{\pi^0 PDG}}{m_{\pi^0 meas}}$. R_{η} is calculated in the same way and the equations in 3.31 and 3.32 can be solved and c_1 and c_2 can be determined:

$$c_2 = \frac{R_{\pi^0}^2 - R_{\eta^0}^2}{\langle E_{\gamma\pi^0 meas} \rangle - \langle E_{\eta\pi^0 meas} \rangle} \qquad (3.32)$$

$$c_1 = R_{\pi^0}^2 - c_2 \cdot \langle E_{\gamma\pi^0 meas} \rangle \qquad (3.33)$$

The resulting values for c_1 and c_2 are then written to the configuration file of the BaF_2-crystals and the next iteration can be performed. As there was not enough η statistics for each crystal in TAPS available, this procedure had to be applied ringwise. Figure 3.18 illustrates the improvements of the accuracy of the energy calibration after a few iterations (this picture has been taken from [32]).

In total six iterations were applied. After the second order correction in energy was accomplished, the center position of the reconstructed η-mass of events (with at least one hit in TAPS) was 547.4 MeV and thus very close to the PDG mass value of the η-meson. Figure 3.19 illustrates this fact.

3.4. BaF_2 Time Calibration

In principle, the time calibration of the BaF_2 modules can be accomplished in the same way as with the NaI channels. To convert a channel number given by the TDC[6] into a time information two parameters are used: the Offset T and the gain factor g. As with the NaI modules the gain factor is given by the TDC hardware. It is a fixed value, which was set to 0.1 ns for all channels. As this value can not be changed, the only parameter that remains to be adjusted is the Offset T. Thus the Offset for each channel needs to be shifted independently in such a manner, that the prompt peaks in the time spectra of all channels are located at the same position (channel). In other words: all

---
[6]Time-to-Digital Converter

3.4. BaF$_2$ Time Calibration

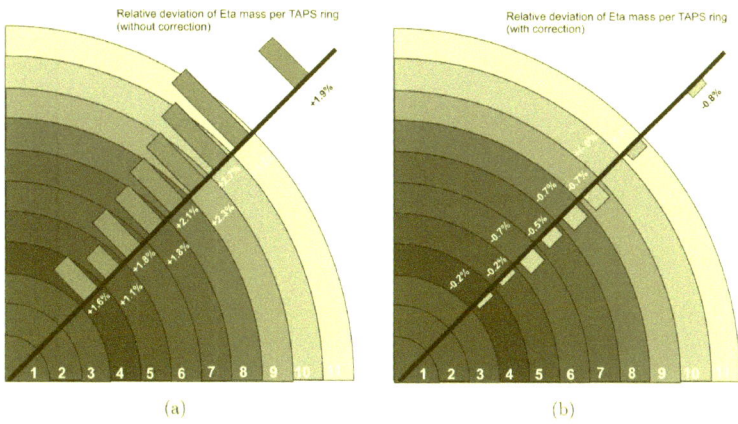

Figure 3.18.: Deviation of the measured η-mass from the PDG value in the case without (a) and with (b) the second order correction.

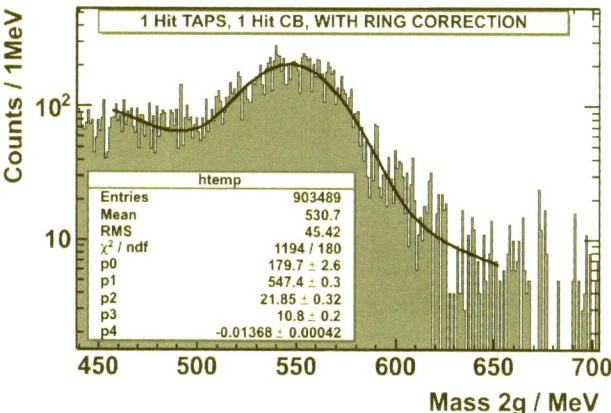

Figure 3.19.: After accomplishment of the second order correction: Reconstructed invraiant mass of $\eta \to \gamma\gamma$ (with TAPS after cuts).

modules have to have the same *time zero*.

3. Calibration

This can be achieved in the following way. Events with two neutral hits in TAPS were selected and analyzed. For every event the detector time information of the first neutral cluster hit t_i and the second neutral cluster hit t_f were read out. The differences $\Delta t_{if} = t_i - t_f$ were stored in a 2D histogram for every BaF_2-channel (the index crystal of the clusters was used, as the index crystal provides the time information of the hit). Thereafter projections for each channel on the time-difference-axis were applied. In all resulting histograms a peak around zero is expected. As the data is not time-calibrated at first, this leads to broad multi-peak structures, due to wrong offset parameters. In order to correct for these, the peaks in all spectra have to be fitted (by a Gaussian). The mean value of this fit is stored as t_{coinc}.

When this is done for the first time, the positions of t_{coinc} are given by wrong and old offset factors, which will be referred to as T_i^0.

$$t_{coinc} = \langle t_i f \rangle = t_i - t_f \tag{3.34}$$
$$= g \cdot (ch_i - T_i^0) - g \cdot (ch_f - T_f^0) \tag{3.35}$$
$$= g \cdot \Delta ch \cdot \Delta T^0 \tag{3.36}$$

If one assumes, that the offset factors were correct (already perfectly calibrated), this would lead to the following situation:

$$t_{coinc} = 0 \tag{3.37}$$
$$0 = g \cdot (ch_i - T_i^n) - g \cdot (ch_f - T_f^n) \tag{3.38}$$
$$0 = g \cdot \Delta ch \cdot \Delta T^n \tag{3.39}$$

As the offset parameters still need to be calibrated, new and corrected factors T_i^n need to be obtained somehow. This can be easily done by subtraction of equation 3.39 from 3.36. Assuming that detector f is already calibrated, this leads to:

$$-\frac{t_{coinc}}{g} = \Delta T^0 - \Delta T^n \tag{3.40}$$
$$= T_i^0 - T_f^0 - (T_i^n - T_j^n) \tag{3.41}$$
$$= T_i^0 - T_i^n - (T_f^0 - T_f^n) \tag{3.42}$$

Based on this assumption $(T_f^0 - T_f^n) = 0$, and thus:

$$\rightarrow T_i^n = T_i^0 + \frac{t_{coinc}}{g} \tag{3.43}$$

Again, this is a procedure which has to be applied iteratively. Hence the offset parameters will converge to their final values and the quality of the calibration depends more or less on the number of iterations. The macro used for this iteration was originally developed by D. Werthmueller, but had to be modified in order to become usable with the newly developed TACalibration class.

The result of the timing calibration of TAPS is shown in Figure 3.20. The FWHM of the fit to the time difference peak from two photon hits in TAPS is 0.66 ns, which is a very good timing resolution compared to prior works [32].

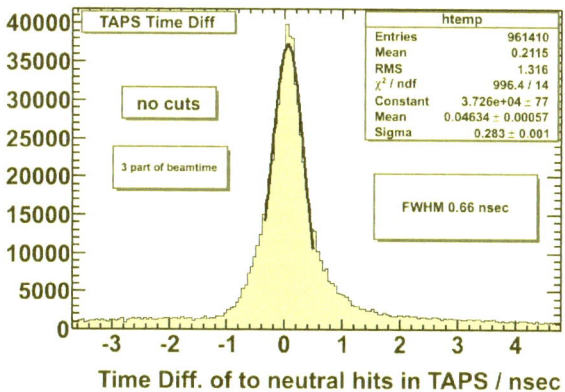

Figure 3.20.: After calibration: time difference of two neutrals detected in the TAPS detector.

3.5. PID ϕ Correlation

As was described in the last chapter, the Particle Identification Detector (PID) is used to detect and identify charged particles. The reconstruction of events works as follows: if there is a hit in the NaI-calorimeter the software checks whether the PID element, which the particle must have passed through, has fired or not. If the element has fired, the energy information dE of this element is read out and the particle is 'marked' as charged. In order to apply this procedure successfully, the PID elements and the NaI-elements have to be correlated correctly in ϕ.

Unfortunately, before the LH_2 beamtimes were started, the PID had been removed and reinstalled in order to install the MWPCs, which were not used afterwards. During this reinstalltion the PID was rotated and thus the original correlation in ϕ was broken. Hence the position of the single PID elements relative to the NaI crystals had to be corrected and the configuration file of the PID elements needed an update. Figure 3.21 illustrates the cases of a correct and a broken ϕ-correlation. As can be easily understood

3. Calibration

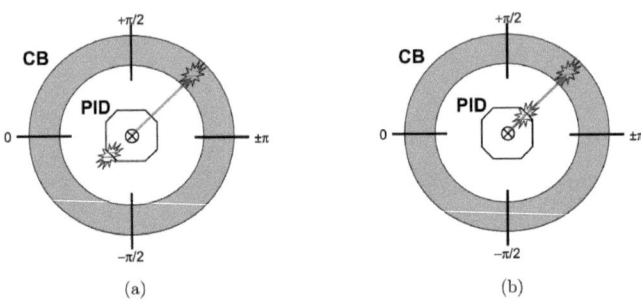

Figure 3.21.: PID: a) the wrong correlation in ϕ leads to false detection of a charged particle as neutral hit. b) PID elements and NaI are aligned correctly. Thus the detection of the charged particle works properly. The picture was taken from [32].

the latter prevents any correct particle identification.

The ϕ-calibration can be done in the following manner:

- In a first step for all coincident hits in the PID and in the NaI clusters, the corresponding PID channel number is plotted in a 2D histogram versus the ϕ angle of the NaI hit (which is given by the geometriy of the Crystal Ball). The result of this is shown in Figure 3.22.
- After that the projections of all 24 PID channels are made and the peaks are fitted by a Gaussian. Thereafter the resulting peak positions (values in ϕ) are used to determine the position of each PID element.
- Based on this, a correct relation between azimuthal ϕ angle and the corresponding PID module can be received and can be written to the PID configuration file.

Figure 3.23 shows the result of this procedure. As can be seen, all PID element were properly corrected in their ϕ position.

3.6. PID Energy Calibration

The PID not only marks charged particles as charged, but it also provides an energy information. As has been described in the previous chapter the amount of deposited energy depends on the particle's mass and energy. Thus protons will deposit more energy than charged pions or electrons and positrons of the same energy. This fact can be exploited by plotting the PID energy versus the energy of the NaI-clusters.

3.6. PID Energy Calibration

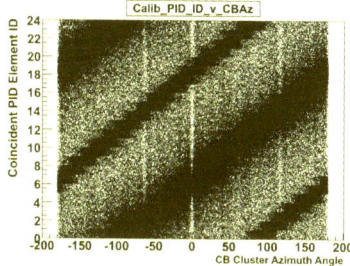
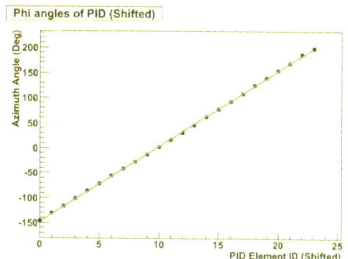

Figure 3.22.: Azimuthal ϕ angle of the NaI hits for coincident hits in the NaI calorimeter and the PID versus the responding PID element ID.

Figure 3.23.: All azimuthal ϕ angles of the NaI calorimeter have been correctly assigned to the corresponding PID elements.

Figure 3.24 shows a 2D histogram with the energy of the PID dE plotted versus the energy of the $Na(Tl)I$ clusters E. Due to their different energy deposit, charged particles can be identified. For this a graphical cut, the so called 'banana-cut', is used. Every charged hit, that is inside such a 'banana-cut' will be assigned to a certain particle ID. In the picture on the right a black proton cut and a blue electron cut are defined. This histogram has been produced in a Monte Carlo simulation.

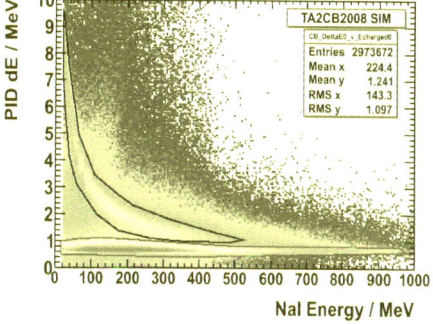

Figure 3.24.: Simulation of $\eta \to e^+e^-\gamma$: PID energy plotted versus the NaI cluster energy. The 'banana bands' are used to identify protons and $e+/e-$.

In fact, the energy of the PID scintillators was not calibrated. Thus it was not possible to define one global proton cut which could be used for all PID elements. The same holds for the e^+e^- and the $\pi^+\pi^-$ 'banana-cuts'.

Instead of performing an energy calibration, independent banana-cuts were defined for each of the 24 PID elements. This solution worked very fine, as will be shown in the chapters concerning the data reconstruction and analysis.

3.7. VETO Energy Calibration

The VETO-modules are used in the same way as the PID detectors; namely to identify charged particles. In earlier experiments the VETO detectors were used just to 'mark' charged particles. Therefore the VETO-bitpattern was read out for every event. VETOs with a digital bit '1' had fired, the ones with '0' were not hit. In 2007 the VETO electronics was upgraded and since then each VETO element provides an energy information. Due to different masses and energies of charged particles, their different energy deposit will lead to distinct band structures similar to the PID (3.6). As each BaF_2 detector has a single $VETO$ module mounted in front, a charged particle will fire the VETO element and the corresponding BaF_2 element(s).

The VETO energy can be plotted against the BaF_2 cluster energy, as is illustrated in Figure 3.25. In this 2D histogram the proton band is clearly visible as is the minimum ionizing bump at $\approx 200 MeV$. The position of the latter depends only on the material and thickness of the VETO elements (section 2.5.3).

But before a single graphical proton 'banana-cut' can be defined and used with all VETO channels, these need to be calibrated.

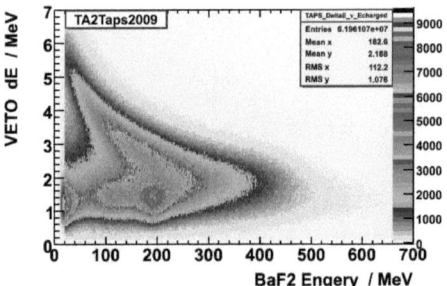

Figure 3.25.: VETO energy plotted versus the BaF_2 cluster energy. Data from the July run in 2007; after calibration.

The procedure of the VETO energy calibration had been developed by T. Gessler [17] in 2007/2008. However, the code provided by T. Gessler had to be modified and adapted to existing programs. The classes TA2Taps2009 and TACalibration of AcquRoot were used to select events of interest and to store them as ntuples in a file. Furthermore, a configuration file containing the VETO-pedestal settings for each element was needed. This was obtained by using a modified version of the TAPS standalone analysis software called AnaPW0[7] and to analyze cosmic data. Thereafter the data in the ntuple file were preprocessed.

For each VETO channel the energy was plotted versus the BAF_2 cluster energy in a 2D histogram. After that all of these were projected onto the VETO energy axis. Then a

---
[7]The original version of the AnaPWO program was developed by P. Drexler; Peter.Drexler@exp2.physik.uni-giessen.de

3.8. VETO Correlation and Time Calibration

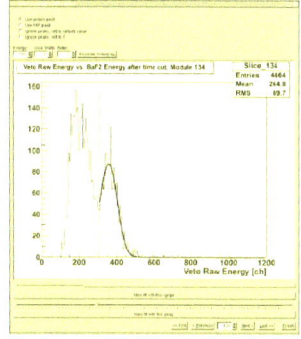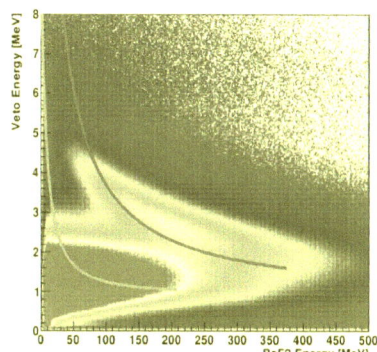

Figure 3.26.: Left: Fit of the proton distributions of all 384 VETO elements. Right: A 2D histogram showing the Veto dE versus the BaF_2 energy. The red and the green curve are theoretical Bethe Bloch calculations and are used in the calibration process.

fitting routine was used to fit the proton distribution for a certain energy range. As the theoretical values had been calculated before (Figure 3.26 right side), the parameters for the energy-to-channel relation in the VETO's configuration file could be corrected. Thus a proper calibration of the VETO was achieved. The beamtimes of July and June in 2007 were the first beamtimes of the A2-Collaboration, in which the VETO energy was calibrated and used for data analysis.

3.8. VETO Correlation and Time Calibration

The time calibration of the VETO channels can be done in a manner similar to the corresponding calibration of the NaI and BaF_2 detectors. In principle the offset parameter of each VETO element has to be corrected in such a way, that the prompt peaks in the timing spectra for all channels are at the same position. This was done. Again the TA2Calibration class was used to store events of interest to a file and after that a fitting routine was used to determine the peak positions.

Another important item is the VETO-channel to BaF_2-channel correlation. As mentioned before, the VETO electronics had been upgraded. During this process some VETO-channels were mixed up, and thus the identification of charged particles did not work properly for all channels. Hence, a correct VETO to BaF_2 correlation had to be established. Fortunately the tools for this procedure were already available, because D. Werthmüller from the Basel group had done this before during a beamtime in 2006.

3. Calibration

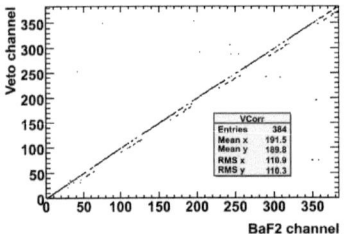
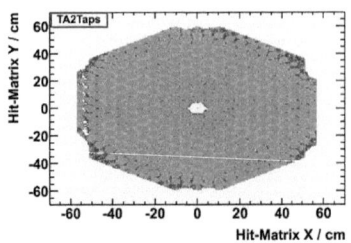

Figure 3.27.: VETO channels plotted versus the BaF_2 channels. Some VETOs are not properly correlated and thus are not on the diagonal line.

Figure 3.28.: The so called HitMatrix of TAPS for charged particles, helps to identify mixed up VETO channels. The shown spectrum does not contain any mixed up channels any more (DATA).

Very helpful in the search for mixed up VETO-Channels is the so called HitMatrix of TAPS for charged particles. In this 2D histogram the hits on the TAPS front-surface are drawn in a normal x and y coordinate system. The intensity is given by the color code. A mixed up VETO-channel in this HitMatrix would be easily discovered, due to wrong intensity distributions (in color - or even 'wholes'). The Figure 3.28 displays the HitMatrix of TAPS for data of the June beamtime from 2007 after the errors in the VETO to BaF_2 channel correlation had been corrected.

3.9. TAGGER Energy Calculation

As a matter of fact, the TAGGER energy is not calibrated, but calculated. The energy of the tagged photons is completely derived from the measured final electron E_{e^-} and the initial electron beam energy E_0 as given in equation 2.3. The energy E_{e^-} of the electron that has radiated is derived from the position at which it hits the focal plane of the TAGGER. The correspondence between this position and the electron energy can be calculated, because the magnetic field strength of the TAGGER magnet is well known.

The program used for this calculation is named 'ugal-v1.C' and was written by J.R.M Annand from the University of Glasgow. Figure 3.29 illustrates the results of this calculation. How this program works in detail, is described in [14] and [2].

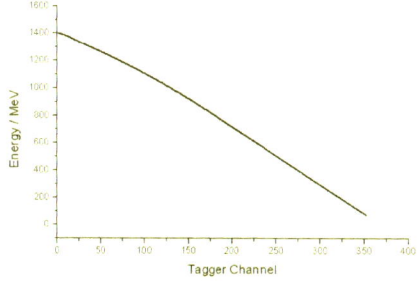

Figure 3.29.: Plot of the calculated TAGGER Photon-Energy against the TAGGER channels.

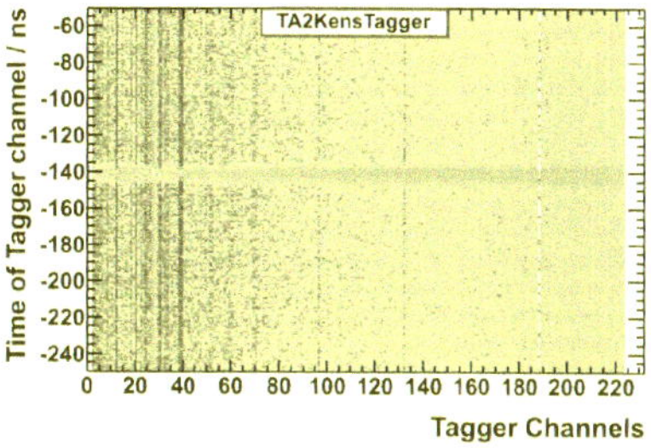

Figure 3.30.: 2D Spectrum plotting the TAGGER time against the TAGGER channels.

3.10. TAGGER Time Calibration

The timing calibration of the TAGGER is a straight forward procedure and thus could be performed very fast. The tools[8] were kindly provided by the Glasgow group. Each channel of the Tagger has a known time conversion of 0.1 ns per channel. This value is based on a calibration, which was done when the TAGGER was first installed. All TAGGER channels need to be aligned in time, so that all 'prompt' peaks in the timing spectra of all channels occur at the same point. The 'prompt' peak is the signal related to the experimental trigger.

The alignment of the channels is done by fitting a Gaussian distribution to the prompt peak of each channel. The mean of this Gaussian is determined and after that a constant time offset is applied in order to shift the mean of each channel to the same arbitrary time. This is important for later analyses. The reason for that is the need to subtract random events, which can easily be done, if the TAGGER has a calibrated timing. In that case a 'prompt' and a 'random' window in the timing spectrum of the TAGGER (all channels combined) can be defined. A more detailed description of this will be given in chapter 4.

3.11. Readout of the TAGGER Scalers

During the beamtimes the raw number of scalers was read out every 20 seconds and was written on the data files. Figure 3.31 shows a scaler readout from the July beamtime in 2007. Each TAGGER channel number corresponds to a certain electron energy and

---

[8] Beyond others this refers to a macro called AlignTDC.C

3. Calibration

thus to a certain photon energy. This transformation is fixed (section 3.9) and does not have to be changed/corrected as long as the magnetic field and the electron beam energy stay the same. Looking at Figure 3.31, one can see that the distribution of the scalers is proportional to $1/E_\gamma$. The distribution is not absolutely smooth, which is related to two aspects. On the one hand a few TAGGER channels were noisy, which led to higher counts rates (e.g channels 29 and 30). On the other hand, some channels were broken or did not work properly (small efficiency), which explains the drops in the scaler counts (e.g channels 39 and 40).

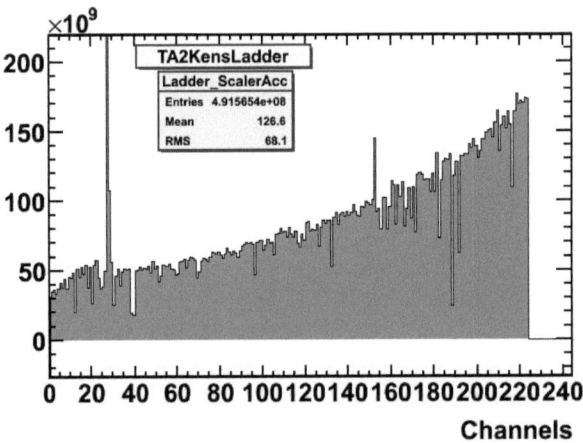

Figure 3.31.: Electron hits in each Tagger channel. The channel number can be converted into a photon energy. Channel 0 corresponds to the highest photon energy.

During the two beamtimes no energies lower than 600 MeV were tagged; the corresponding segments of the TAGGER had been switched off. The explanation for this is, that due to the $1/E_\gamma$ distribution the lower photon energies would have caused too high count rates in the TAGGER. Furthermore, as the purpose of this beamtime was the production of η, ω and η', the lower beam energies (below 600 MeV) were physically uninteresting.

Before the photon flux can be calculated the scaler rates have to be corrected. Whilst the acquisition of experimental data, the detector systems (CB, TAGGER, TAPS) have *dead times*. During these *dead times* it is not possible to detected an event and store new data, since the system is still occupied with processing the last event. The fraction of time a system is nod dead is referred to as *lifetime*. The TAGGER has the longest *lifetime* compared to the Crystal Ball and TAPS. As a consequence not all detected hits in the TAGGER (scalers) produced events that could be detected in the CB and TAPS.

Thus the scalers need to be corrected using a correction factor c_{sc}:

$$c_{sc} = \frac{\Gamma_{DAQ}}{\Gamma_{TAGGER}} = \frac{\frac{S_1}{S_0}}{\frac{S_{145}}{S_{144}}} \qquad (3.44)$$

Where Γ_{DAQ} is the *lifetime* of the data acquisition of the whole system (CB, TAPS, TAGGER) and Γ_{TAGGER} is the pure *lifetime* of the TAGGER. The normal method to determine these quantities is to calculate the fractions of the lifetime pulsers (S_1 for the DAQ and S_{145} for the TAGGER) and the free running pulsers (S_0 for the DAQ and S_{144} for the TAGGER). As these scalers suffer from a possible overflow, the addition of an integer number is required (meanwhile this problem has been fixed).

The scaler correction factors were supposed to be the same for both beamtimes (June/July 2007) as the same triggers, target, beam current and energy were used. Unfortunately it was found that the scalers of the June beamtime were broken; that means, the scalers were not correctly written on the data files and thus could not be read out.

The determined dead time correction factor is (for the July 2007 beamtime):

$$c_{SC-07/2007} = 0.8101 \qquad (3.45)$$

3.12. The Tagging Efficiency Measurement

The radiated photons pass through a collimator. Thus the raw number of electrons N_{e^-} hitting the focal plane of the TAGGER does not match the number of photons N_γ impinging on the target. Hence, the fraction of the photon beam intensity that gets lost in the collimator has to be determined. Therefore, a correction factor called *tagging efficiency* ϵ_i has to be applied on the number of detected electrons:

$$N_\gamma^i = N_{e^-}^i \cdot \epsilon_i \qquad (3.46)$$

In this equation the index i refers to the channel number of the focal plane element and thereby also to a certain energy. For the determination of the tagging efficiency factors ϵ_i special data runs were performed during the beamtime periods. For these runs a very low electron beam current was used and a special lead glass detector (as described in section 2.2.1). During these special runs the whole DAQ, except for the TAGGER, was switched off. Further on special background runs were performed; for these the beam was switched off. The final calculation of ϵ_i was done by using a macro (TaggEff.C) written by J.R.M Annand. The result of this calculations is shown in Figure 3.32.

Compared to the determined *tagging efficiencies* of former MAMI-B experiments (max. electron beam energy of 850 MeV) the tagging efficiency of this work went up from

≈ 40% [2] to ≈ 70%. The reason for this is the increased electron beam energy of 1508 MeV. The opening angle of the bremsstrahlung beam is given by (for relativistic electrons)

$$\Theta \approx \frac{m_{e^-}}{E_{e^-}} \qquad (3.47)$$

That means, that for a higher electron beam energy, the produced photon beam is already better collimated before it reaches the collimator. Thus the smaller loss results in a higher *tagging efficiency*.

For the July beamtime[9] in 2007 the average *tagging efficiency* was determined as

$$0.6799 \;\pm\; 0.008 \;\text{(with background correction)} \qquad (3.48)$$
$$0.6582 \;\pm\; 0.007 \;\text{(without background correction)} \qquad (3.49)$$

3.13. The Photon Flux

In the sections before (3.12) and (3.11) the correction factors for the *tagging efficiency* ϵ_i and the dead time c_{SC} have been determined for every channel of the TAGGER. Based on this, the photon flux can be calculated. For each TAGGER channel i the number of produced photons in the flux N_γ^i is given by

$$N_\gamma^i = N_{e^-}^i \cdot c_{SC} \cdot \epsilon_i \qquad (3.50)$$

with N_e^i indicating the number of electron hits, which is given by the scaler. The total flux can be obtained by an integration over the range of interest. Table 3.3 lists these numbers for different energy ranges (only July 2007 beamtime). These results were obtained after an analysis of 227 data files. The complete photon flux for the whole energy range (617 MeV to 1400 MeV) is $1.25 \cdot 10^{13}$. The noisy channels were corrected[10]. Using the dead time correction factor (3.45) leads to

$$\text{Corrected Photon Flux } = 1.01 \cdot 10^{13} \qquad (3.51)$$

For the photon flux per data file follows:

$$\text{Corrected Photon Flux } = 4,657 \cdot 10^{10} \text{ per file} \qquad (3.52)$$

---

[9]In the June beamtime 2007 the scalers were broken and could not be read out. Hence, a determination of the photon flux for this beamtime was not possible. Therefore the tagging efficiency for this beamtime was not determined.

[10]This correction was accomplished in the following way: if a channel was noisy, the value of the scaler of the neighbouring channel was used.

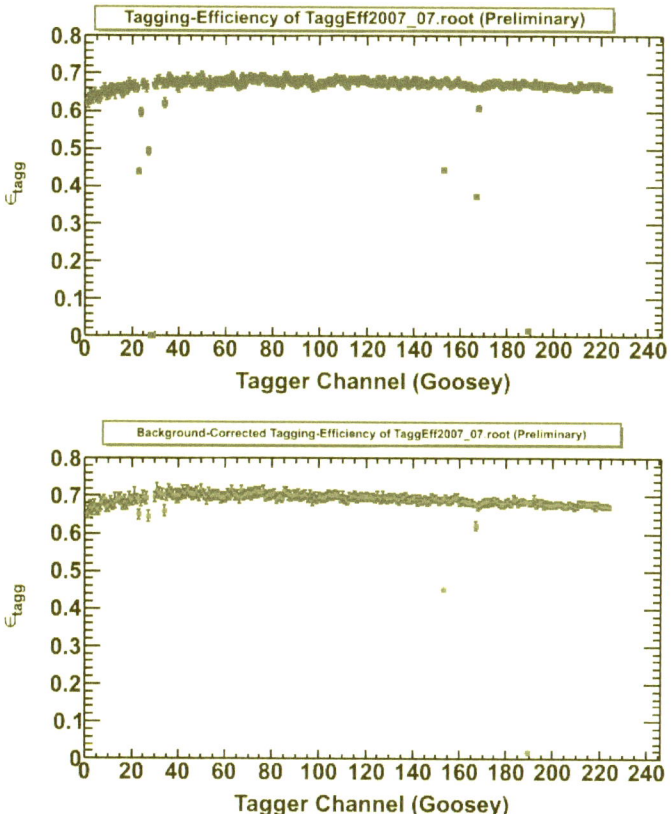

Figure 3.32.: Tagging efficiency for each TAGGER channel (top) and with background correction (bottom).

As for the η-Dalitz analysis not the whole energy range was used (but only from 750 MeV to 1210 MeV) a slightly different flux had to be calculated (Table 3.4). Correcting the 'sum' given in the table by the dead time correction factor leads to:

$$\text{Effective Photon Flux} = 5.88 \cdot 10^{12} \text{ for } \eta \to e^+e^-\gamma \qquad (3.53)$$
$$\text{Flux per File} = 2.620177448 \cdot 10^{10} \qquad (3.54)$$

3. Calibration

Energy range [MeV]	Photon Flux	Comment
617-700	$2.12 \cdot 10^{12}$	
700-800	$1.92 \cdot 10^{12}$	
800-900	$1.82 \cdot 10^{12}$	
900-950	$8.63 \cdot 10^{11}$	
950-1000	$7.68 \cdot 10^{11}$	
1000-1050	$6.96 \cdot 10^{11}$	
1050-1100	$7.21 \cdot 10^{11}$	
1100-1150	$6.93 \cdot 10^{11}$	
1150-1200	$6.78 \cdot 10^{11}$	
1200-1250	$6.03 \cdot 10^{11}$	
1250-1300	$6.01 \cdot 10^{11}$	
1300-1350	$6.08 \cdot 10^{11}$	Nosy channel corrected 28
1350-1400	$5.07 \cdot 10^{11}$	
Sum:	$1.25 \cdot 10^{13}$	
	$1.01 \cdot 10^{13}$	dead-time corrected
935-985	$8.77 \cdot 10^{11}$	
985-1035	$7.67 \cdot 10^{11}$	
1035-1085	$7.27 \cdot 10^{11}$	
1085-1135	$6.66 \cdot 10^{11}$	
1135-1185	$6.60 \cdot 10^{11}$	
1185-1235	$6.28 \cdot 10^{11}$	
1235-1285	$6.11 \cdot 10^{11}$	
1285-1335	$5.81 \cdot 10^{11}$	
1335-1385	$4.93 \cdot 10^{11}$	
1385-rest	$1.97 \cdot 10^{11}$	
Sum	$6.21 \cdot 10^{12}$	
	$5.03 \cdot 10^{12}$	dead-time corrected

Table 3.3.: Photon flux determination of beamtime 07/2007 for certain energy ranges (without dead-time-correction). The broken channels were corrected - values of neighbours were used.

3.14. Verification of the Energy Calibration

Whether a calibration in energy of the calorimeters (NaI,BaF_2) is correct or not, does not only depend on the center position of the mass peak of the η and π^0 meson averaged over all momenta. An important requirement is that for every momentum range the calibration is still appropriate, meaning, that for every momentum range the centers of the invariant (η,π^0)-mass peaks have to be close to their corresponding PDG values. This can be checked easily. The Figures 3.33 and 3.34 show the invariant mass plotted against the momentum of the η-meson. In fact the performed energy calibration guar-

3.14. Verification of the Energy Calibration

Energy Range [MeV]	Photon Flux
750-800	$9.57 \cdot 10^{11}$
800-900	$18.27 \cdot 10^{11}$
900-950	$8.63 \cdot 10^{11}$
950-1000	$7.68 \cdot 10^{11}$
1000-1050	$6.96 \cdot 10^{11}$
1050-1100	$7.21 \cdot 10^{11}$
1100-1150	$6.93 \cdot 10^{11}$
1150-1210	$8.14 \cdot 10^{11}$
Complete	Sum
750-1210	$7.26 \cdot 10^{12}$

Table 3.4.: Photon flux without dead time correction for the η-Dalitz analysis (beam-time 07/2007).

antees a sufficient stability in momentum, as can be seen in the figures. The left hand figure is the result of a simulation of $\eta \to e^+e^-\gamma$ without kinematic cuts. The right hand figure shows the situation in case of the experimental data; here the bands stemming from $\eta \to \gamma\gamma$ and $\pi^0 \to \gamma\gamma$ can clearly be identified. Moreover these histograms show, that the invariant mass is stable for all momenta.

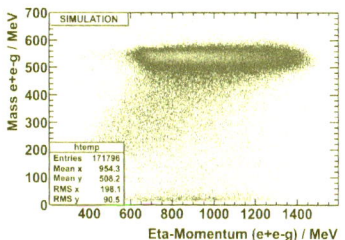

Figure 3.33.: Mass$_{e^+e^-\gamma}$ plotted versus Momentum$_{e^+e^-\gamma}$ for simulated $\eta \to e^+e^-\gamma$ events.

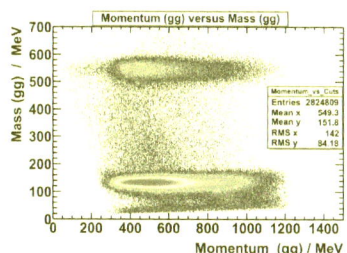

Figure 3.34.: Analysis of $\gamma\gamma$ events of experimental data.

In a further step projections onto the invariant mass axis were produced for 100 MeV wide slices in the whole momentum range. The π^0 and η peaks in each resulting histogram were fitted in order to determine the exact positions of the peaks. The result of this procedure is shown in the Figures 3.35 and 3.36. The maximal deviation from the PDG mass is 1.48% for π^0 and 0.7% in case of the η-meson.

Hence, the performed energy calibration can be considered as accurate and correct.

3. Calibration

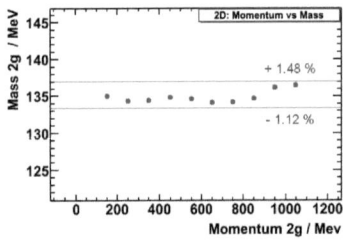

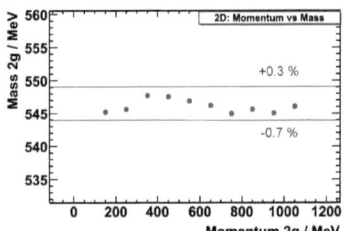

Figure 3.35.: Plot of the reconstructed invariant π^0-mass versus the momentum of the π^0.

Figure 3.36.: Plot of the reconstructed invariant η-mass versus the η-momentum.

The histogram in Figure 3.37 shows the invariant mass spectrum of $\gamma\gamma$ events for the η-mass range (including hits from the TAPS and the CB). This result has been achieved after analyzing one part of the data and after application of cuts on the prompt peaks in the timing spectra. A fit to the η mass clearly demonstrates the accuracy of the overall-energy-calibration (reconstructed η mass = 547.9 MeV).

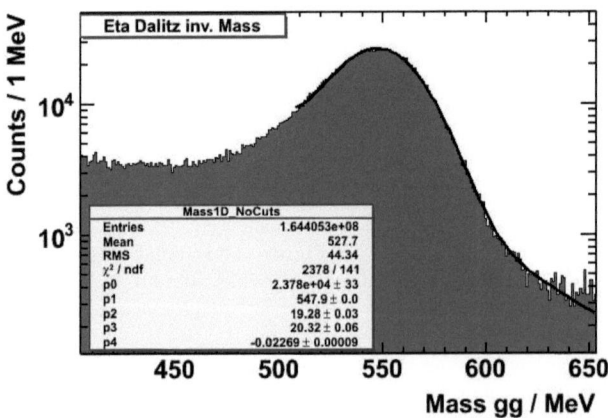

Figure 3.37.: Invariant η-mass after cuts. Experimental data from TAPS and CB.

4. Particle identification and event reconstruction

It is a complex task, to convert the data from a series of stored digital pulse heights and timing signals to a physical interpretation of hadronic interactions. This task requires much thought, knowhow, experience and effort. In this chapter the general chain of data reconstruction will be explained and the used programs will be described.

4.1. Software

The software package is split into several parts. During data taking the DAQ-part of the program AcquRoot is used to combine and store all information from each detector system to disk. For the simulation of the experiment a Monte Carlo program named A2sim is used, which is based on GEANT4. To investigate the experimental data as well as the simulated data, the analysis functions in the user-part of the AcquRoot (section 4.1) program are used.

4.1.1. The AcquRoot Analyzer

AcquRoot is a C^{++} program designed to take and analyze A2-data and was developed by J.R.M Annand of the University of Glasgow. It consists of two main parts: the AcquRootSystem and the AcquRootUser. The former contains all the important software classes for data decoding/encoding and controlling the data-flow of each detector component. This is the basic part of the whole analysis software and at the same time, it is the biggest and most complex part. Thus it should not be changed nor altered by anyone other than the developer himself.

As the experiment is permanently changed or upgraded[1], the AcquRootSystem needs to be maintained constantly; in other words updated and improved.

---

[1] This refers to changes of the electronics and of course to the installation of new hardware, such as the $PbWO_4$ modules.

4. Particle identification and event reconstruction

Due to the software updates, several versions of the AcquRoot-System are available, which are all compatible to the realized experiments from 2006 until today (2010). These are version $4v0$, $4v1$, $4v2$, $4v3$ (and $4v4$ which is being developed at the moment). For the A2-Giessen-Software framework the versions $4v0$, $4v1$ and $4v2$ were used. Later a modified version $4v2+$ has been developed.

The AcquRootSystem is installed separately and is independent from the version of the AcquRootUser, which will be described in the following subsection.

This separation shall assure, that the 'normal users' only change code in the AcquRootUser, but never in the system part. The AcquRootUser provides the environment for developing analysis functions.
Figure 4.1 illustrates the general structure of AcquRoot.

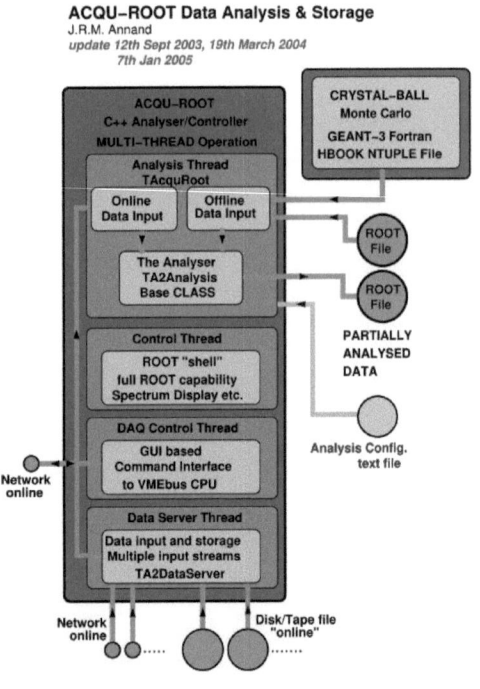

Figure 4.1.: The AcquRoot(System) data storage and analysis system. Extracted from [3].

In Giessen, the A2-analysis group first used the AcquRootSystem as a separate program, as designed; but as new calibration techniques[2] were developed, which led to different setups of the configuration-files, the en/decoding functions of the system software had to be modified. Furthermore installing two independent software parts ends with trying to convince both parts to work together with each other, which often can result in an enormous effort. Thus the decision was made in Giessen, to develop one single version of AcquRoot, that contains an appropriately modified AcquSystem and the AcquUser part for data analyses. The result of this is a program called $AR_{HB}2v3$.

---

[2]Non-linear channel-to-energy relations for the calorimeters, new *time-walk*-functions etc.

4.1. Software

Standard AcquRootUser

The version 4.0 of the AcquRootUser software released by J.R.M Annand is compatible to the AcquRootSeystem $4vX$ versions. This program provides an environment for the analysis of physical reactions. In the class TA2PhotoPhysics[3] all the information provided by the TA2Apparatus[4] classes are collected and prepared for an analysis. The latter means, that the Lorentz vectors and timing information of the detected particles are retrieved and sorted into arrays. These functions are already available and normally should not be changed by the user.

What the user has to change and/or code are special functions to analyze the created arrays of Lorentz vectors. In principle the analysis procedure works as follows: assume a single $\gamma\gamma$-event is analyzed. Then a function is needed, that calculates the sum of two Lorentz vectors. Furthermore the invariant mass as well as the missing mass can be calculated and cuts can be applied. Thereafter, if the event has survived the cuts, histograms can be filled.

The disadvantage of this procedure is, that whenever a cut is changed, the program needs to be recompiled and the whole amount of raw-data has to be analyzed again, which leads to an enormous waste of time, since the run of the software over all data takes approximately up to two days under normal conditions.

Another big disadvantage of the TA2PhotoPhysics is, that only the Lorentz vectors and the timing information of each hit are available in this analysis class. If more information[5] about a detected particle would be of importance in order to properly analyze the event, this would not be possible with the TA2PhotoPhysics class and the normal user would have to modify the detector classes himself, in order to let these provide the desired information to the analysis class.
Because of these two major disadvantages, a new and appropriate classes had to be developed matching the following important requirements.

- All information about a 'hit' have to be available in the analysis classes.
- Changing cuts should not require a recompilation of the main analysis program and a reanalysis of all the raw-data. Thus an event selection should be applied before the actual analysis is performed. These selected events should be saved to ntupl-files, which can be analyzed in less time (up to 3 hours).

The development of a program, that fulfills these requirements, was started in 2007 and took over two years. Meanwhile the program is a fully functional A2-data-Analyzer

---
[3] The TA2PhotoPhysics is the standard AcquRoot class for data analysis and is derived from the TA2Analysis class.
[4] The TA2Apparatus is the basic class for the detector systems; the TA2Tagger, TA2KensTagger, TA2CrystalBall and the TA2TAPS are derived from this basic class.
[5] i.e: cluster size, dE, pure cluster energy, short gate energy, etc.

4. Particle identification and event reconstruction

named $AR_{HB}2v3$ and has been tested throughly.

$AR_{HB}2v3$ Giessen standard Analyzer

During the development of $AR_{HB}2v3$ former versions $AR_{HB}0v1$, $AR_{HB}1v0$, $AR_{HB}2v0$, $AR_{HB}2v1$ and $AR_{HB}2v2$ were released. $AR_{HB}2v3$ is a standalone program containing its own AcquRootSystem based on $4v2$, but with important modification and thus it will be referred to as $4v2+$. These modifications were important and needed in order to be able and use a non-linear channel-to-energy transformation for the NaI and $BaF2$ calorimeters as well as for the use of the new *time-walk* functions.

The main improvements compared to the standard AcquRootUser $4v0$ lie on the side of the application of analysis procedures and routines.

The first major aspect during the development was to make all information about detected 'hits' available in the users analysis classes. Thus a special class named TA2HenningsParticles had been introduced, which was constantly improved and adjusted to the requirements of a proper analysis. When this class was finally finished, it was renamed into TA2Particle[6]. Furthermore a new class for the TAPS detector was developed, which was given the name TA2Taps. The existing classes TA2CrystalBall and TA2KensTagger were modified; the former was renamed into TA2CB2008.

The second aspect was to introduce an analysis based on ntuples. As with introduction of the TA2Particle class a completely new analysis procedure had to be developed anyway, the old class TA2PhotoPhysics had to be replaced. The analysis part of the $AR_{HB}2v3$ is based on three new classes, which are the TA2PhotoReconstruction, the TA2PhotoBasicAnalysis and the TA2PhotoAnalysis.

Figure 4.2 provides a schematic overview on the structure of the $AR_{HB}2v3$-program. For the VETO the class TA2Veto is used, for the PbClass the TA2TaggPbGlass, for the PID the TA2PlasticPID, for the NaI the TA2CalArray, for the $BaF2$ the TA2TAPS_BAF2 and for the MWPCs the TA2CylMWPC.

In the following the main classes concerning the analysis will be listed and described:

- The TA2Particle is used in the TA2Apparatus class to store all information about a single hit; such as the Lorentz Vector, the cluster size, cluster energy, charged or not, dE-values, short gate information, detector timings, etc. The C++ classes for the big detector systems TAPS, CB and TAGGER are all based on the TA2Apparatus class. Thus they can fill arrays of TA2Partcile objects and provide these to the main analysis class (TA2PhotoReconstruction). The TA2Particle is not a stand

---

[6]As the idea of a TA2Partcile class was adopted by other people (S. Schumann), there are meanwhile different versions of this class available.

4.1. Software

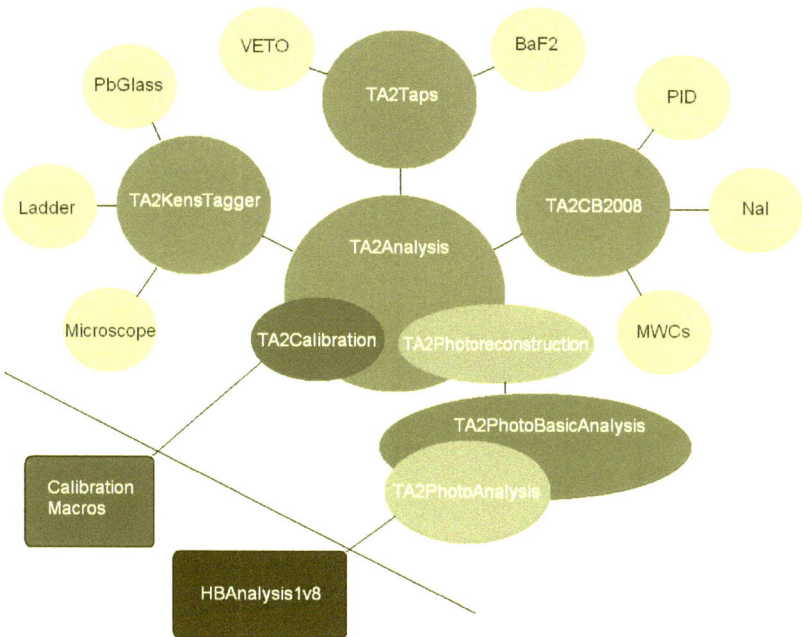

Figure 4.2.: A schematic representation of the principle structure of the $AR_{HB}2vX$ program line. The output of a $AR_{HB}2vX$ program can be further analyzed or used by calibration macros or the HBAnalysis1v8.

alone class, but it is derived from the TA2ESDParticle, which will be described in 4.3.4.

- The TA2PhotoReconstruction is the basic class for all further analysis steps. In principle, it communicates with the detector systems. Therefore it contains objects of the corresponding apparati (TA2CB2008, TA2Taps, TA2KensTagger). For every event it retrieves all detected particles as objects of TA2Particle from these apparati. In the next step, it will sort these particles into a photon-array, proton-array and so on. During this procedure it learns the number of each sort of particles within the actual event. Finally it generates a particle-set based on the structure called ParticleData, which contains all sorted particles-arrays and numbers. This particle-set is then given to the TA2PhotoAnalysis and the TA2PhotoBasicAnalysis. It should be mentioned, that different sets of particles for each event can be generated and given to the analysis class. The idea of this

4. Particle identification and event reconstruction

is the fact, that sometimes a particle identification accomplished by an apparatus class is not 100 % accurate. In such a case the apparatus will assign a second particle-ID to the deteced hit, which can be used in the TA2PhotoReconstruction to create a second/alternative set of detected particles which shall also be analyzed.

- The TA2PhotoBasicAnalysis is, as its name implies, the class containing the basic analysis functions for inclusive and exclusive investigations of $\eta \rightarrow 3\pi^0\gamma$, $\omega \rightarrow \pi^0\gamma$ and $\pi^0/\eta \rightarrow \gamma\gamma$. For the latter an onlyTAPS, onlyCB and combined analysis function is available. As every installation of $AR_{HB}2v3$ comes with this class, any user can at once accomplish a first and quick analysis. Furthermore users can compare their results more easily, because the analysis functions are the same. As the TA2PhotoBasicAnalysis is not derived from any other $AR_{HB}2v3$-class, and only uses the particle-set-structure to receive information, it is possible to create a shared library and use this class in other programs (even for other experiments). As the purpose of this class is to check data-files and/or the results of a calibration, the reconstructed physical information (e.g invariant mass) are stored in histograms, which can be looked at while running. Furthermore it is possible to create ntuples.

- The TA2PhotoAnalysis is derived from the TA2PhotoBasicAnalysis. Thus, when using the TA2PhotoAnalysis (which is the standard setting in the configuration file of the 'PhotoAnalysis') all basic functions are still available. By setting a '1' flag in the configuration file, these basic functions will be executed ('0' = off). The purpose of the TA2PhotoAnalysis is, to give all users an environment and the possibility to define their own analysis functions. Thus, after a new installation of $AR_{HB}2v3$, this class will not contain any analysis function. In principle the idea is, to create and use functions in this class to analyze events and save all acquired information on a ntuple in a root-file (which can by further analyzed using the HBAnalsis1v8). This method leads to an enormous reduction of the data in size and thus helps to save a lot of time, because analyzing the few ntuple-files, that have been created, can be accomplished rather fast.

- The TA2Calibration is used to prepare any calibration. As all calibration macros need a certain set of information (stored in 1D, 2D or even 3D histograms), these need to be extracted from the raw data, preprocessed and prepared in the right manner and stored. This is done by the functions and routines of the TA2Calibration, which can be activated easily by switching a certain flag to '1' in the configuration file of the 'PhotoAnalysis'.

- The TA2CreatESD is a class, that can be used to create and store an '**E**vent **S**ummery **D**ata' (**ESD**). How this is done, and how the ESD can be analyzed later, will be described in 4.1.3.

Other important and new classes of $AR_{HB}2v3$ are:

- TA2Taps is the new TAPS class. It is derived from the TA2Apparatus, that contains all basic functions for every detector system. Due to some upgrades of the TAPS hardware and electronics it became necessary to develop a new class, because the old class[7] did not work anymore with the new/improved setup. As the VETOs after the upgrade also provided a dE information, the $dEvE$ method of particle identification became accessible (section 4.3.2). Furthermore the *time-of-flight* method and the *pulse-shape-analysis* can be used. As it may occur, that a particle identification is not clear, a second-ID (introduced in the TA2Particle) can be used to save this 'information'. This has never been done before in the A2-experiment.

- The TA2CB2008 is based on the TA2CrystalBall written by J.R.M Annand. In order to implement the usage of the TA2Particle and the second-IDs, as with TA2Taps, some modification had to be applied and thus the resulting class was renamed.

- The TA2KensTagger (modified) was only slightly modified in order to enable the use of the TA2Particle class. Thus the name was not changed.

- TA2Veto is the new class for the VETOS. Since after the VETO-upgrade, these detectors deliver also energy information. Thus a new treatment of the VETO-readouts in the software led to the development of a new class.

Concerning the analyses accomplished by the author himself (chapter 6) special functions had to be developed in the TA2PhotoAnalysis. Each of these functions was for analyzing a certain class of events (e.g: one proton, two γ and two charged). All important variables (e.g missing mass etc.) were calculated/investigated and saved independently to different ntuples. In this way a separate set of ntuple-files was created for each analysis and thus a 'data-compression' was performed.

4.1.2. The HBAnalysis1v8 NTuple Analyzer

The HBAnalysis1v8 program is a pure ntuple-analyzer. It can open any root-file and read contained ntuples, and thus it can be used to analyze all sorts of ntuple data from any experiment in the world. The user has to edit just one file named 'HBAnalysis.C'. In this, the output-file and all input-root files have to be specified, as well as the 'JOBs' and the 'CUTs'. A 'JOB' refers to the following procedure:

'Read all entries of a ntuple-variable, apply cuts and plot a histogram'.

As the user may want to apply more than just one cut, or even wants to apply many and different sets of cuts, certain 'JOB'-types need to be available (and of course can be defined).

The number of input-files is not limited. The maximal number of different JOB-types

---
[7]TA2TAPS written by R. Gregor.

4. Particle identification and event reconstruction

is 30; each 'JOB'-type may contain 100 jobs and thus one instance of this analyzer can produce 3000 histograms. For each 'JOB'-type a 'CUT'-list can be specified with a maximal number of 40 individual cuts. But not all cuts are of the same type, as will be described in the following:

- The AddCut: this adds a simple cut to the 'CUT'-list (e.g a cut on the π^0-mass would be simple cut).

- The AddORCut: assume a cut shall be applied on the coplanarity, which would mean, that a value can only survive this ORcut if it is in (e.g) the range from $-180 \rightarrow -160$ OR from $160 \rightarrow 180$.

- The AddGCut: it is common to use graphical cuts based on the TCutG class of ROOT and apply these on 2D histograms. These cuts can be used as well in the HBAnalysis1v8.

- The AddSpecialGCut: assume a 2D graphical banana cut shall be applied on the 'time-of-flight' information of protons in TAPS. Therefore, a flag needs to be set in another variable, telling the program that it IS a TAPS hit. Otherwise the cut will not be applied.

This tool provides a very comfortable way of analyzing data. This is demonstrated with the following example: the number of tested cut-settings in the analysis of the η-Dalitz decay is in total 67. If this would have been done not using a ntuple-based analysis procedure, but the standard AcquRootUser analysis procedure, consequently the whole amount of raw data would have had to be analyzed 67 times; wheres each time would have lasted 2 days (in total $\approx$ 140 days). If one uses instead the ntuple-analyzer on the preselected ntuple files, one cut-setting can be tested in less than 30 minutes.

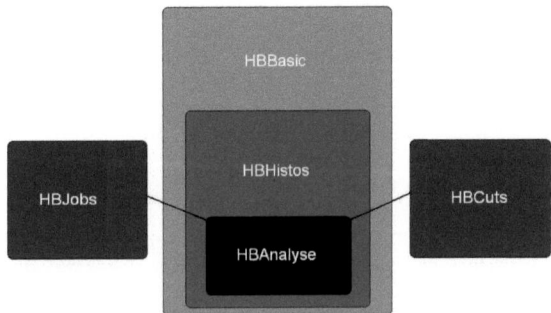

Figure 4.3.: Schematic illustration of the structure of th HBAnalysis1v8 program.

4.1. Software

HBAnalysis1v8 consists of 5 main classes, which have been constantly updated and improved. Figure 4.3 illustrates the principle structure of the program HBAnalysis1v8. The author plans to increase the functionality of this program to the analysis of 'Event Summary Data'.

4.1.3. The Event Summary Data Project

The idea of an 'Event Summary Data' (**ESD**) is simply to create out of raw data a set of files only containing reconstructed events. Furthermore generating an ESD would mean to compress the data at the same time.

When analyzing any A2-Data using an AcquRoot program, raw data are read and being decoded using calibrated settings in the configuration files. Every hit itself in one of the detector systems is being reconstructed, the particle is identified and then turned over to the analysis class, where all those information from the detectors are combined to an 'event' and thus are prepared for the actual analysis steps, which will follow. All this work could be reduced to just the 'following analysis steps'.

Let us assume a set of experimental data was already calibrated properly by experts. Then the basic steps of decoding and reconstruction could be done just a single time writing the combined 'events' one by another to a 'root'-file (ESD). Thereafter, whenever one wants to analyze data, one has only to copy the ESD of interest, open the files and analyze the pure 'events'.

To provide such a comfortable manner of analysis, some effort and thought has been necessary. In case of the A2-experiments, an upgrade of the AcquRoot has to be developed, that enables the program to write an ESD on files. Therefore one main aspect is the development of an 'event class'. During a research visit at the Niels Bohr Institute, the author developed[8] an ESD-event class called 'TA2ESDEvent'. The simple idea is, to use objects of this class as a container, in which 'particles' of 'a single event' can be stored. As the original 'TA2Particle' class was found to be too large in bit-size, another rather small class named TA2ESDParticle had to be developed as well. All what had to be done in addition, was to create a class for AcquRoot, that enabled this program to use the new 'event and particle' classes and to write an ESD to disk. The class developed for this purpose is called TA2CreateESD and can simply be used with AcquRoot by replacing the 'TA2PhotoReconstruction' by 'TA2CreateESD' in the main configuration file. Thus, AcquRoot will not execute any analysis functions, but will combine all detected particles event wise and store them to files.

The new classes have been developed for $AR_{HB}2v3$ and were tested thoroughly. It

---

[8]This could only by achieved based on the intense support by Christian Holm Christensen. http://www.nbi.ku.dk/english/staff/beskrivelse/?id=168679 and http://cholm.web.cern.ch/cholm/

4. Particle identification and event reconstruction

should be mentioned that the TA2ESDParticle and the TA2ESDEvent (together with other ESD-reader related classes) are independent of the main AcquRoot program. Thus a shared library can be created and used with every other experiment as well.

Once the ESD has been created, a program is needed to open and read the ESD-files. In order to prove that the whole concept really works, the author developed a small macro called 'ESDreader.C', which can read any ESD, that uses the TA2ESDevent and TA2ESDParticle classes. As was mentioned before, the newly invented analysis classes for $AR_{HB}2v3$ (TA2PhotoBAsicAnalysis and TA2PotoAnalysis) can also be turned into independent shared libraries and the purpose of this is, to be able and use these analysis classes in the ESD-Reader.C. This has been tested and works fine. Still, the ESD-Reader.C has rather to be considered a prototype which does not provide a complete environment for fast analyses. This is, because it is only a root-macro and thus it does not work very fast. Furthermore it has no multi-threading and thus one can not look at histograms as these are being filled. The next step would be, to develop a fully independent stand-alone program with multi threading (and maybe a graphical interface).

4.1.4. The A2-Sim Monte Carlo Simulation

The A2-Simulation was developed by D. Glazier[9] and is based on GEANT-4 (c++). The program A2-Sim contains the experimental setup as shown in Figures 4.4 and 4.5. For the simulation of physical events and the interaction of particles with the detector material the functions provided by the GEANT package are used. These so-called Monte-Carlo simulations of physical reactions are important to determine the detector response and the acceptance for every decay-channel of interest.

The GEANT-4 version installed and used in Giessen is 4.9.0. Two different versions of the A2-Sim program were used: A2.06.05.08 and A2.23.10.09. For the former a small modification of the setup was applied, namely the implementation of a Nb-target, which was not available in the A2-Sim and thus could not be used before. Beyond other analyses, the Giessen group analyzes in-medium effects of the ω-meson. In this respect two Nb-beamtimes are upon investigation, and as the detection efficiencies have to be determined by using the A2-Sim program, the Nb-target needed to be implemented.

More information about the A2-Sim program can be found in [18], as a very detailed description of the latest version of this program is available in the corresponding manual.

---

[9]Derek Glazier from the University of Edinburgh, dglazier@ph.ed.ac.uk .

4.1. Software

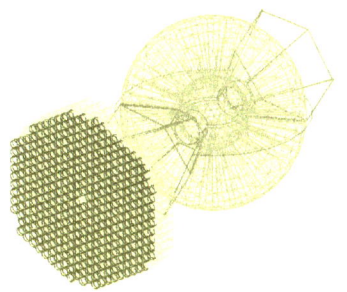

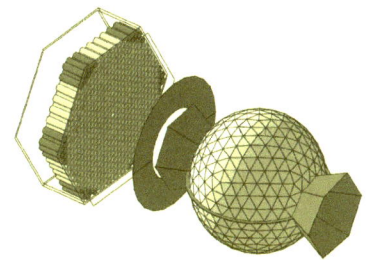

Figure 4.4.: Detector geometry implemented in GEANT-4.

Figure 4.5.: The CB/TAPS detector systems in the GEANT geometry of the A2-Sim program.

Phase Space Event Generator

If one wants to simulate a certain reaction using A2-Sim, one has to provide a starting distribution, which can be generated by using an appropriate event generator. For former analyses in Giessen GEANT-3 Monte-Carlo simulations had been performed based on an old FORTRAN event generator called 'evgen-brems'. The original version of this event generator was written by D. Hornidge[10]. Evgen-Brems only provides phase space distributions, and thus it can not be used in order to determine a realistic acceptance correction in the case of Dalitz decays. Since no other event generator was operational for some time, 'evgen-brems' was even used to determine a preliminary acceptance correction for the η-Dalitz decay. When later the PLUTO event generator became available, a correct simulation of the Dalitz decays of η, ω and π^0 was performed. In chapter 5 a comparison between the results of the different η-Dalitz simulations is discussed.

Other reactions like $\eta \rightarrow \gamma\gamma$ or $\omega \rightarrow \pi^0\gamma$ can be simulated using phase space distributions. The only difficulty with the old event generator was, that the output of this old FORTRAN tool could not just be used as input for the new A2-Simulation program based on GEANT-4 and ROOT. Thus a converter-program needed to be installed and used.

The phase space event generator is a simple program, that mainly consists of a random number generator which produces arbitrary values for angles and momenta in a given range for chosen particles. The A2-Simulation expects direction unit vectors in x, y, z, momentum, and energy; but the event generator only provided ϕ and θ in radians.

---

[10]http://www.mta.ca/ dhornidg/ .

4. Particle identification and event reconstruction

Hence, a transformation had to be implemented:

$$P_x = \sin(\theta) \cdot \cos(\phi) \tag{4.1}$$
$$P_y = \sin(\theta) \cdot \sin(\phi) \tag{4.2}$$
$$P_z = \cos(\theta) \tag{4.3}$$
$$E = \sqrt{P_{tot}^2 - m_{Particle}^2} \tag{4.4}$$

For all simulations a $1/E_\gamma$ distribution for the energy of the incident photons was used and only quasi free production was assumed. Later, the Fermi motion effect was implemented into the generator; this is important especially when simulating reactions induced on heavy target nuclei (e.g Pb, Nb, Ca etc.), because all nucleons have momenta in some direction even if the nucleus is at rest.

The center of mass energy is calculated depending on the mass of the nucleon (proton) and the energy of the incident photon.

$$s = m_N^2 + 2 \cdot E_\gamma \cdot m_N \tag{4.5}$$
$$w = \sqrt{s} \geq 2 \cdot m_{Meson} + m_N \tag{4.6}$$

If this energy is larger or equal to the production threshold of the desired reaction the appropriate GEANT functions[11] are executed. The return values of these functions are four momenta and angles in the center of mass system (**CM**). As GEANT-4 requires those variables in the laboratory system (LAB), they have to be transformed. This can be achieved using the following equations[12]:

$$|\vec{p}_{CM}| = \sqrt{(p_{CM}^{4th}) - m_{Meson,\,p}^2} \tag{4.7}$$

$$\theta_{CM} = \arccos\left(\frac{p_{CM}^{3rd}}{|\vec{p}_{CM}|}\right) \tag{4.8}$$

$$\phi_{CM} = \arctan\left(\frac{p_{CM}^{2nd}}{p_{CM}^{1st}}\right) \tag{4.9}$$

$$\beta = \frac{E_{Photon}}{E_{Photon} + m_{Meson,\,p}} \tag{4.10}$$

$$\gamma = \frac{1}{\sqrt{1-\beta^2}} \tag{4.11}$$

$$p_{LAB}^{4th} = \gamma \cdot p_{CM}^{4th} + \beta\gamma p_{CM}^{3rd} \tag{4.12}$$
$$p_{LAB}^{3rd} = \gamma \cdot p_{CM}^{3rd} + \beta\gamma p_{CM}^{4th} \tag{4.13}$$
$$p_{LAB}^{2nd} = p^{2nd} \tag{4.14}$$
$$p_{LAB}^{1st} = p^{1st} \tag{4.15}$$

---
[11] These functions refer to the two body reaction function GDECA2 and the three body function GDECA3.
[12] The four components of each vector are marked with: $1st, 2nd, 3rd$ and $4th$.

4.1. Software

These values are calculated for each particle of the reactions generated by 'evgen-brems'. Thereafter all values are saved to a ntuple on a hbook-file, which has to be transformed into a root file, as was explained before.
GEANT-4 reads event-by-event from this root-files and does all the tracking of the particles as well as the calculations of energy deposit and loss. Thereby it generates an output in a certain format which later can be decoded by the AcquRoot analyzer.

The PLUTO Eventgenerator

The PLUTO is an advanced event generator which is capable of calculating Dalitz distributions. Hence, it was used for a realistic simulation of the Dalitz decays under investigation in order to determine an exact detector acceptance.

In simple terms, PLUTO is a ROOT-based framework for implementing customized event generators. Its main features are:

- It is object-oriented (C++), modular, flexible and extendable.

- Enables fast simulations of kinematics and decays.

- Filters (e.g. acceptances, efficiencies)

- It does not provide transport calculations through media (nor geometry, nor field)!

Pluto was designed by M. Kagarlis[13] at the GSI in 2000/2001. The code used for this work was kindly provided by I. Fröhlich[14].

Unfortunately using PLUTO for the analyses in Giessen was not simple, as the output of this event generator was not compatible to the program 'A2-Sim' and thus could not be used as 'input' without further ado. After some work was put into developing an own converter, which was never finished, an already working converter became available from the TAPS group of the University of Basel[15]. Thus the simulations based on the PLUTO starting distributions could be performed without further delay.

All accomplished simulations and their analyses will be described in chapters 5 and 6.

---

[13] Marios Kagarlis, http://www-hades.gsi.de/computing/pluto/html/PlutoIndex.html
[14] Ingo Fröhlich, http://webdb.gsi.de/hades_webdb/hades_collab.hc_hades_homepage.people_info?p_id=75
[15] Special thanks at this point to Manuel Dieterle, who did all the simulation-converter work; and thanks to Dominik Werthmüller, who wrote the converter and made this work possible.

4. Particle identification and event reconstruction

4.2. Hardware

Analyzing and calibrating a beamtime requires a certain computer infrastructure. As in Giessen five beamtimes were being investigated and thus had first to be calibrated, new servers had to be installed and a new infrastructure was set up.

In order to match these requirements certain facts had to be taken into account:

- How much disc space is required to store the data of five (or more) beamtimes?
- How many CPUs are needed to realize the calibration and analyses in a reasonable time frame?
- What would be an appropriate user management, and how could the management of the software be realized?
- What kind of computer-infrastructure is necessary, appropriate and affordable.

The data of a single beamtime amount on average to roughly 1000 GByte. As the data of former beamtimes from 2003 and 2004 were already stored on the only two servers, filling up most of their disc space, the decision was made, that a total disc capacity of 22.000 GByte would be appropriate and match all (future) requirements in that respect. Another important fact is, that the storage-systems have to be fast and have to guarantee a high availability of the data. Furthermore the possibility of a loss of data needs to be very small. Matching these requirements two Raid5, two Raid6, one Raid0, and one Raid1[16] systems were installed.

Concerning the estimation of the number of required CPUs a simple calculation can help. A modern CPU (Intel 2.4 GHz core) needs approximately 450 seconds for analyzing one data file, which has a standard size of 2 GByte. As larger beamtimes consists of up to 500 files, this leads to a runtime larger than 60 hours. To calibrate all components of all detectors for a complete beamtime the programs need to run over the data up to 50 times, due to the iteration methods of the calibration techniques and due to the fact, that very often the program crashes because of damaged files, which need to be sorted out. This has to be done for each beamtime. Hence, the data needs to be split into several parts of equal size and several instances of the programs have to be used at the same time. Using 4 CPUs per beamtime would decrease the runtime by a factor 4. As additional calibration work for other A2-Collaboration partners had to be done[17], and besides this calibration work Monte-Carlo simulation were planed to be performed, a total number of 24 CPUs was installed. In 2009 the number of CPUs was increased to 30.

---

[16]RAID stands for a 'redundant array of independent disks'. For further information please consult standard literature concerning this topic.
[17]This refers to the following two beamtimes: 2008_11_He3 and 2008_12_Pi0 .

4.2. Hardware

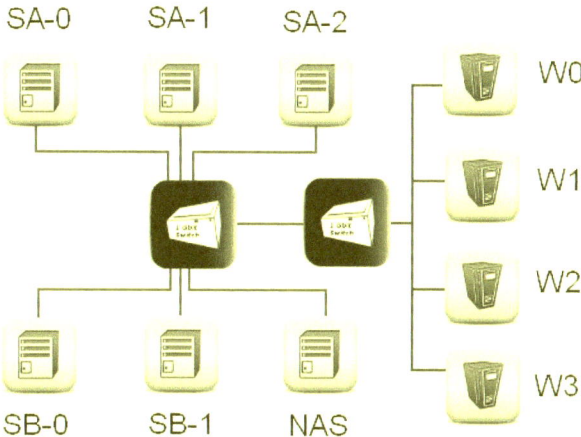

Figure 4.6.: Server/Computer infrastructure of the A2-Group of the University of Giessen. The blue lines and the black boxes are the 1GBit Ethernet network. See Table 4.1.

	Name	Services	CPUs	RAM [GB]	RAID [TB]	OS
SA0	schneewittchen	NIS, NFS	4	4	6	Suse 10.2
SA1	sensemann	NFS	8	8	10+2	Suse 10.3
SA2	zwergenchef	NFS, BackUp	2	4	2+2	Suse 9.3
SB0	piggy	-	4	4	-	Suse 10.2
SB1	snickers	-	4	4	-	Suse 10.3
NAS	nas01	NFS	1	0.5	1(2)	Linux
W0	goofy	NFS	4	4	2	Suse 11.1
W1	dino	-	2	2	-	Suse 11.1
W2	pi0	-	1	0.5	-	Suse 10.2
W3	honigbrumsel	-	2	2	-	Suse 10.2

Table 4.1.: List of servers and workstations used for calibrations and analyses.

It was an important aim to have equal user and user-homes on each server/workstation. This was realized by introducing a NIS network. NIS stands for Network Information Service which is an information system for managing networks and provides beyond other options a network based user management. Thus all user-IDs and user-home directo-

4. Particle identification and event reconstruction

ries setup on the NIS-server are distributed to all other servers/workstations allowed to connect. Hence, a user can work and start programs on all machines in the NIS-network. Figure 4.6 illustrates the setup of the Giessen server/computer infrastructure. The server SA0 is running the NIS.

As a user is now able to start any program (e.g AcquRoot) on each machine, but the data are stored on a certain server, a special network based file-management has to be established. This can be realized using the NFS service. NFS stands for Network File System, which is a server-protocol allowing a user on a client computer to access files over a network in a manner similar to how local storage is accessed.

For this reason the professional servers (SA0, SA1, SA2 in Figure 4.6) which have the large Raid systems installed, run each a NFS server-service. All secondary servers (SB0, SB1) as well as the workstations in the office rooms (W0, W1, W2, W3) have access to the NFS and thus can mount the data directories.

The free CONDOR cluster software was installed. The idea was at first to use a cluster managing software to control analysis and calibration jobs. It was found that these methods have a major drawback. As the installed network is an 1 GBit network based on copper-Ethernet, allowing realistic transfer rates up to $\approx$ 35 MBytes per second and per link, only a certain amount of data can be transfered from the NFS-servers to the secondary servers and workstations, which were also always used for calibration and analysis tasks. As a modern CPU can process up to 5 MBytes of A2-data per second, the transfer rate only supports enough bandwidth for 7 external CPUs. Hence, a cluster-management service would not increase the speed of data analyses because the software does not know how to distribute jobs under optimal exploitation of the bandwidth (at least in the tried/examined setup using CONDOR).

A different and more straight forward solution was found. In a first step the programs (AcquRoot, A2-sim) were prepared for simultaneous usage. As the maximum number of people in the A2-Group in Giessen actively using the servers/workstations for analyses was only three, the machines were assigned to the users in such a manner that the transfer of the data to the CPUs did not exceed the bandwidths.

4.3. Analysis Procedure

An analysis has to be optimized concerning the run time. Furthermore each step in a procedure must be comprehensible and revisable, otherwise mistakes are not traceable and unclarities can not be solved. The analyses in Giessen are accomplished in a four step process.

- Particle reconstruction in the detector apparati classes in AcquRoot.

4.3. Analysis Procedure

- Collecting and combing all detected particles from the detector objects (TA2Taps, TA2CB2008, TA2KensTagger) in the TA2PhotoReconstruction class, where thereby 'events' are created.

- Preselection of (interesting) events. Calculation of all interesting values (missing mass, coplanarity etc.). Saving of these information in an ntuple-file.

- Analyzing the ntuple-file (event wise) under application of cuts on the variables, that had been saved to the nutple before.

These steps are described in more detail in the following sections.

4.3.1. Particle Reconstruction in CB

As the MWPCs were not read out during the liquid Hydrogen Beamtimes in July and June 2007, the only detector components of the Crystal Ball usable for reconstruction of particles were the NaI-calorimeter and the PID detector, which have been described in section 2.4.1 and section 2.4.2.
To identify a particle the information of the **P**article **I**dentification **D**etector (PID) is used. For every hit in the NaI-calorimeter the correlated channel of the PID is checked, whether it has fired or not. Since uncharged particles like photons and neutrons do not fire the PID this method provides the means to separate charged hits from uncharged hits.

An uncharged hit will always be assigned to the photon-ID. The reason for this is, that in the CB no techniques are available to separate photons from neutrons. The only method that can help to reduce the neutron background is a narrow cut on the prompt peak in the timing spectrum.

For the separation of charged particles into protons, charged pions or electrons/positrons the PID detector is used. As has been described in section 3.6 the identification of particles is done using the $dEversusE$ method, which means that the energy of the PID (dE) is plotted versus the energy of the NaI-cluster E. Figure 4.7 shows such a 2D plot for the PID channel 23. For each[18] channel of the PID one proton cut, one π^+/π^--cut and one e^+/e^--cut was defined. All these cuts are so called 'banana'-cuts using the class 'TCutG' of the ROOT framework. As overlapping banana-cuts caused to many problems, they were re-defined in an appropriate form without overlapping. Thus either the e^+/e^- or the π^+/π^- can be used, as these particles populate almost the same area of the 2D plot. A particle that can not be identified will be treated as 'rootino'. Table 4.2 illustrates the LOGIC of the particle identification in the Crystal Ball. '1' stands for fired and fulfilled, '0' for not fired and not fulfilled.

---

[18] As was described in the PID calibration section, the energy of the PID was not calibrated. Thus individual 'banana'-bands were defined for each of the 24 channels.

4. Particle identification and event reconstruction

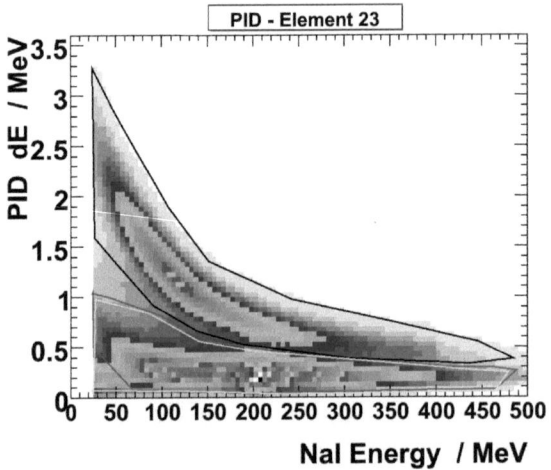

Figure 4.7.: Data: 2D-plot showing the energy of PID channel 23 versus the energy of the NaI calorimeter. Furthermore a graphical proton cut (black), π^+/π^--cut (red) as well es an e^+/e^--cut can be seen.

NaI fired	PID fired	Proton-Cut	e^+/e^--cut	π^+/π^--cut	ID
1	0	0	0	0	photon
1	1	0	0	0	rootino
1	1	1	0	0	proton
1	1	0	1	0	electron
1	1	0	0	1	piplus

Table 4.2.: LOGIC for particle identification in the CB apparatus.

4.3.2. Particle Reconstruction in TAPS

Due to the upgrade of the VETO electronics in 2006/2007, the vetos ever since provide an energy information. Thus the dE versus E method is applicable, as with the PID and CB. Furthermore, the larger distance of the TAPS system to the target provides an advantage for the identification of particles. As this distance is large enough the *time-of-flight* method can be used too.

Moreover the BaF_2 modules provide the possibility to perform a *pulse-shape-analysis* (PSA), which is an appropriate technique to identify nucleons and thus to identify neutrons. The PSA was not used for this experiment, for the following reasons:

4.3. Analysis Procedure

BaF_2 fired	Veto fired	$dEvE$ cut			TOF cut	ID
		p	π^+/π^-	e^+/e^-	p	
1	0	0	0	0	0/-	γ
1	0	0	0	0	1	γ
1	1	1	0	0	-	p
1	1	0	1	0	-	$\pi^{+/-}$
1	1	0	0	1	-	$e^{+/-}$
1	1	0	0	0	-	rootino
1	1	1	0	0	1	p
1	1	1	0	0	0	$\pi^{+/-}$
1	1	0	1	0	0	$\pi^{+/-}$
1	1	0	0	1	0	$e^{+/-}$
1	1	0	0	0	1	rootino
1	1	0	0	0	0	rootino

Table 4.3.: LOGIC of the TAPS (TA2Taps) particle identification. '-' stands for 'not active', '1' means fulfilled, '0' stands for 'not fulfilled'.

- Only exclusive photo nuclear reactions off the proton were under investigation.
- Background stemming from 'neutron-hits' can be removed without a PSA. This is because of the larger TAPS-to-target distance and the resulting 'later timings'.
- The short gate component of the BaF_2 (which is used for the PSA) was not calibrated.

Different to the CB, the new class of TAPS supports overlapping 'Banana'-cuts. This is, because two main techniques ($dEvE$, TOF) can be used together and furthermore a 'second-ID' can be assigned to each particle. Table 4.3 illustrates the LOGIC of the TA2Taps[19] particle identification. This table contains only the used methods and banana-cuts. The assignment of secondary IDs was not used.

Figure 4.8 illustrates a 2D time-of-flight histogram and a proton banana cut. Figure 4.9 illustrates a similar 2D histogram plotting dE against E.

4.3.3. Reconstruction and Separation of e^+e^- from $\pi^+\pi^-$

Before an event can be reconstructed, all detected particles have to be collected from the apparati objects (of TAGGER, TAPS, CB) in the software. This is done by the class TA2PhotoReconstruction. Special routines of this class determine the number of each particle type within each event. Furthermore a certain function sorts the particles by

---
[19]TA2Taps is the C++ class for the TAPS apparatus.

4. Particle identification and event reconstruction

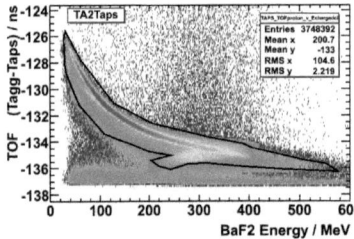
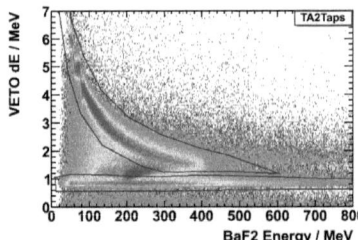

Figure 4.8.: From simulation: a 2D histogram plotting the *time of flight* (TAGGER-Time minus TAPS-Time) against the BaF_2 energy. Also the 2D proton *tof-cut* is shown (black line).

Figure 4.9.: From simulation: a 2D histogram plotting the VETO energy against the BaF_2 cluster energy. A proton band as well as an $e+/e-$ can be seen.

their particle-ID and saves them to separate particle-arrays. In the next step these arrays and the numbers of each particle sort are filled into what is called 'a particle data set'. This set is then given to the analysis class TA2PhotoAnalysis/TA2PhotoBasiAnalysis.

The main difficulty in the analysis of Dalitz decays using the CB/TAPS setup in Mainz is, that no magnetic field is available and thus it is difficult to accomplish an exact separation of light particles that are charged ($e^{+/-}$, $\mu^{+/-}$, $\pi^{+/-}$). As most charged Myons are not stopped by the electromagnetic calorimeters (NaI, BaF_2), the major task is to separate electrons/positrons from charged pions. As without magnetic field the sign $(+/-)$ of the charged hits can not be determined, electrons and positrons will be referred to as electrons, and charged pions will always be called piplus (this way it is handled in the software too).

How this separation can be realized shall be explained with the example of the following two decays of the η-meson:

$$\eta \rightarrow e^+e^-\gamma \qquad (4.16)$$
$$\eta \rightarrow \pi^+\pi^-\pi^0 \qquad (4.17)$$

If the aim is to analyze the decay 4.16, then all charged hits (non-proton), will be reconstructed as electrons. This is realized by using an 'electron-banana'-cut in the $dEvE$-identification procedure. Then all real charged pions will incorrectly be reconstructed as electrons too, and thus the wrong mass will be assigned to these particles. This has a strong effect on the kinematics. As in the analysis the momentum balance in x, y, z is checked as well the energy balance, these false 'electrons' (in reality charged pions) are supposed not to fulfill the cuts. Hence, the charged pion background can be eliminated by this technique (as will be proofed in the following).

4.3. Analysis Procedure

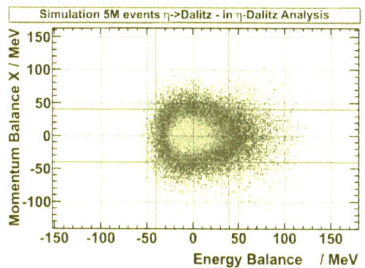
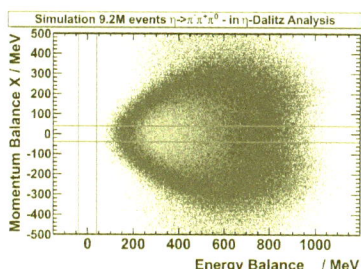

Figure 4.10.: Simulation of η-Dalitz events. 2D-Plot of the *momentum balance in X* versus the *energy balance*.

Figure 4.11.: Simulation of $\eta \to \pi^+\pi^-\pi^0$ in η-Dalitz analysis. 2D-Plot of the *momentum balance in X* versus the *energy balance* (before cuts).

A closer look on the variables *momentum balance* and *energy balance* shows, that the cuts which are applied on the kinematics in the η-Dalitz analysis (chapter 6) very strongly suppress background from charged pions. This fact is illustrated by the two 2D-plots shown in the Figures 4.10 and 4.11. In the first the momentum balance in X direction is plotted versus the energy balance for true η-Dalitz events[20]; the cuts that are applied on these variables in the analysis (chapter 6) are illustrated by the orange (red) lines. In the latter Figure the corresponding variables are plotted in case of a simulation of $\eta \to \pi^0\pi^+\pi^-$ analyzed in the η-Dalitz analysis-function of $AR_{HB}2v3$. Hence, the suppression of background from charged pions is successful under application of the described means. The suppression factor has been determined as $F_{suppress} \approx 3 \cdot 10^{-7}$ (section 6.4, equation 6.34).

Further on cuts on the *missing mass* help to suppress and eliminate background from charged pions[21]. In Figure 4.12 the missing mass (of a proton) is plotted for simulated $\eta \to \pi^0\pi^+\pi^-$ events in the η-Dalitz analysis. As the range of the applied cut is 910 − 980 MeV nearly all events are removed.

Furthermore electrons generate larger clusters in the calorimeters than charged pion do. Figure 4.13 and Figure 4.14 illustrate this fact. This fact can be exploited too, by cutting on the cluster size and by analyzing only those clusters, that exceed the specified minimum cluster size. The Figures 4.15 and 4.16 illustrate the success of this method. Both figures show the result of the same simulation of 9.2 million events $\eta \to \pi^+\pi^-\pi^0$. In the former, the number of detected events is shown without any cuts applied (all charged

---

[20]From a MC-simulation of 5 million events.
[21]Again this works only because the wrong mass (m_e) was assigned to the $\pi^{+/-}$.

4. Particle identification and event reconstruction

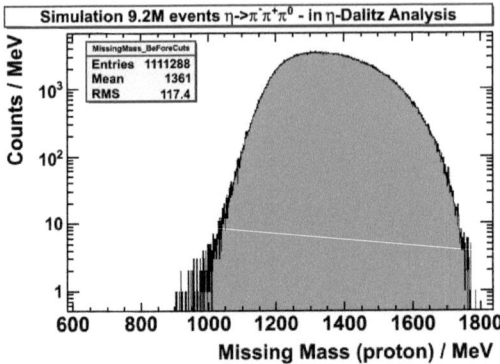

Figure 4.12.: Simulation of $\eta \to \pi^+\pi^-\pi^0$ in η-Dalitz analysis before cuts. Plot of the *missing mass*.

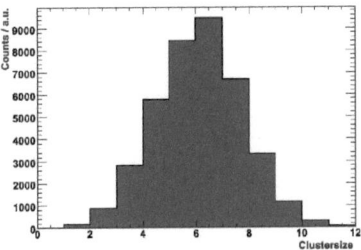

Figure 4.13.: From simulation: Cluster size of detected electrons/positrons. This picture has been taken from [6].

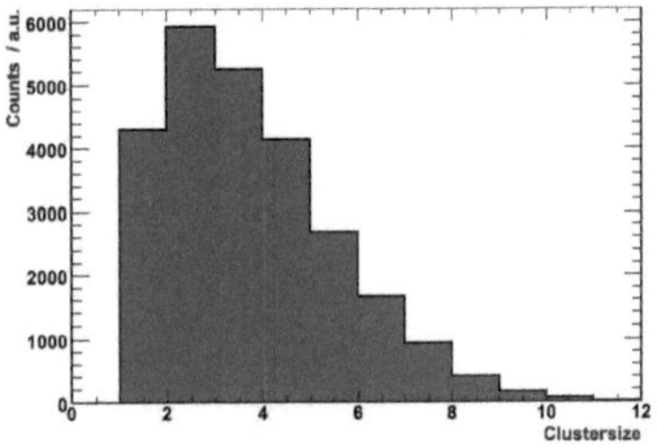

Figure 4.14.: From simulation: Cluster size of detected π^+/π^-. This picture has been taken from [6].

pions were reconstructed as 'electrons'). The latter figure shows the same plot after all cuts on the kinematics and the cluster sizes; as can be deduced from this histogram: no events survive. In section 6.4 all background channels in the η-Dalitz analysis (including channels with $\pi^+\pi^-$) are discussed.

Figure A.18 (appendix) shows a 2D-plot of the momentum balance in X direction versus the energy balance in the analysis of $\eta \to \pi^+\pi^-\pi^0$ in the experimental data. As can be seen from this 2D-plot, the cuts on the kinematics are fulfilled, when the charged hits are reconstructed correctly as $\pi^+\pi^-$ (compare to Figure 4.11).

4.3. Analysis Procedure

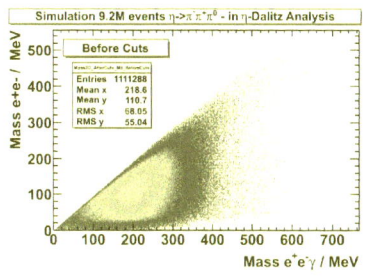 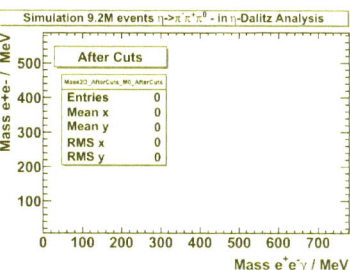

Figure 4.15.: Simulation: $\eta \to \pi^+\pi^-\pi^0$ in the η-Dalitz analysis before cuts.

Figure 4.16.: Simulation: $\eta \to \pi^+\pi^-\pi^0$ in the η-Dalitz analysis after cuts.

4.3.4. Event Selection and Data Compression

In the analysis the following decays and reactions were investigated:

$$\text{Dalitz decays}:$$
$$\eta \to e^+e^-\gamma \qquad (4.18)$$
$$\pi^0 \to e^+e^-\gamma \qquad (4.19)$$
$$\omega \to e^+e^-\pi^0 \qquad (4.20)$$
$$\text{Other charged decays}:$$
$$\eta \to \pi^+\pi^-\pi^0 \qquad (4.21)$$
$$\omega \to \pi^+\pi^-\pi^0 \qquad (4.22)$$
$$\text{Neutral decays and reactions}:$$
$$\pi^0 \to \gamma\gamma \qquad (4.23)$$
$$\eta \to \gamma\gamma \qquad (4.24)$$
$$\eta \to \pi^0\gamma\gamma \qquad (4.25)$$
$$\eta \to \pi^0\pi^0\pi^0 \qquad (4.26)$$
$$\omega \to \pi^0\gamma \qquad (4.27)$$
$$\pi^0\pi^0 \; - \; \text{Production} \qquad (4.28)$$
$$\pi^0\eta \; - \; \text{Production} \qquad (4.29)$$

All these decays and reactions were investigated exclusively, i.e., besides the listed final state particles (4.18-4.29) the detection of a proton was required. Thus only photon induced reactions off the proton were investigated[22].

---
[22] A proton target was used (lH_2).

4. Particle identification and event reconstruction

For each decay-channel an independent set of ntuple-files was created for a further, separate analysis. This was done in the analysis class TA2PhotonAnalysis in which special routines were used to calculate all interesting values (such as missing mass, coplanarity) for each decay-channel separately. All this information was written on the ntuple-files. How these variables were determined shall be explained by a simple example. Assuming an event $\eta \to \gamma\gamma$ is being analyzed exclusively (proton required). In this case the momentum balance $\vec{P}$ can be calculated in the following way (all vectors are Lorentz 4-vectors):

$$\underbrace{p_{\text{target}} + p_{\gamma-\text{beam}}}_{input} = \underbrace{p_{\gamma 1} + p_{\gamma 2}}_{\eta} + p_{\text{recoil}} \qquad (4.30)$$

$$P = 0 = p_{\text{target}} + p_{\gamma-\text{beam}} - (p_{\gamma 1} + p_{\gamma 2} + p_{\text{recoil}}) \qquad (4.31)$$

Each component of P can be retrieved by using a certain operation of the TLorentzVector class of ROOT: e.g. $P_z = P.Pz()$. In a similar manner the missing mass, coplanarity, energy balance, and so forth can be calculated.

It has to be mentioned, that for all analyses the 'higher multiplicities' were investigated too; i.e., if there were more particles detected than the expected number of different particles corresponding to a certain final state, all possible combinations among the particles resulting in a valid final state were built and analyzed.

4.3.5. Application of Cuts

The generated nutple-files can be analyzed using the program HBAnalysis1v8 (section 4.1.2). This tool provides the possibility to apply cuts on the variables (e.g. missing mass, timing etc.) contained in the ntuples. Certain cuts have already been applied during the phase of the particle reconstruction in AR_HB2v3 within the apparati classes (section 4.3.1 , section 4.3.2); these are the banana-cuts on the *time-of-flight* and the *dEvE* bands. As the HBAnalysis1v8 supports the class 'TCutG' of ROOT too, the user can choose and apply stricter cuts on these variables again.

Most important are the cuts on the kinematics. The variables of interest are the missing mass, momentum balance in x,y,z, the energy balance and the coplanarity. Furthermore two body simulations (chapter 5) provide helpful information. E.g. from those it can be learned that the maximum Θ-angle of the backscattered proton for an η-production is 50°, if the beam energy is less than 1400 MeV. Information like these can help to apply appropriate cuts on all available variables; thus, the background can be extremely reduced.

In chapter six all accomplished analyses will be described and for each analysis the applied cuts will be listed.

5. Simulation

This chapter gives an overview on all simulations. The performed *two-body-calculations* as well as the start distributions for the Monte Carlo simulations of the Dalitz decays will be discussed. Furthermore a comparison between the results of two η-Dalitz simulations, one using a phase-space distribution and the other using the correct Dalitz distribution, will be presented.

A detailed discussion of the results of all analyses of the simulated data will be discussed in chapter 6.

Table 5.1 lists all performed Monte Carlo simulations[1]. For most of the simulations a phase space start distribution was used; only for the Dalitz simulation a Dalitz-distribution was generated using the PLUTO generator. All programs used are described in chapter 4.

5.1. Two Body Calculations

Simple *two-body* calculations provide an appropriate means to gather helpful information on the basic kinematics of the reactions of interest. Thus the maximum θ-angles of the involved particles, the backscattered proton and the produced meson, can be determined. The calculations were performed for the photo production of the following mesons: π^0, η, η' and ω. As decays of the ω-meson were also investigated by the CB/ELSA-group at the University of Giessen, the two body calculation concerning the ω meson were additionally performed for higher incident energies (up to 3.5 GeV). All calculations are available for download [7].

5.1.1. η-Production Calculations

As for the investigation of the η-Dalitz decay every helpful information in oder to suppress the background was of importance, the kinematics of the η-production off the

---

[1] All simulations were performed by the author himself. The conversion of the PLUTO-output into a file-format readable by the MC-sim program was performed by M. Dieterle of the University of Basel.

5. Simulation

Reaction/Decay	Start Distribution	Number of Events [million]
$\eta \to e^+e^-\gamma$	Phase Space	0.5
$\eta \to e^+e^-\gamma$	Pluto	0.25, 2.5 and 5.0
$\eta \to \pi^+\pi^-\pi^0$	Phase Space	0.9, 1.9 and 9.2
$\eta \to \pi^+\pi^-\gamma$	Phase Space	0.9
$\eta \to \gamma\gamma$	Phase Space	10.0
$\eta \to \pi^0\gamma\gamma$	Phase Space	3.0
$\eta \to \pi^0\pi^0\pi^0$	Phase Space	1.0
$\omega \to e^+e^-\pi^0$	Pluto	5.0
$\omega \to \pi^+\pi^-\pi^0$	Phase Space	0.3
$\omega \to \pi^0\gamma$	Phase Space	2.0
$\pi^0 \to e^+e^-\gamma$	Pluto	5.0
$\pi^0 \to \gamma\gamma$	Phase Space	2.0
$\pi^0\pi^0$ − Production	Phase Space	1.0
$\pi^0\eta$ − Production	Phase Space	3.0(*)
π^0-Production off the neutron	Phase Space	0.5

Table 5.1.: Performed MC-Simulations. (*)In this case ten times 0.3 M events were simulated for (ten) different intervals of incident energy.

proton in a photo-nuclear reaction had to be calculated. The reaction given by

$$\gamma_{\text{beam}} + p_{\text{target}} \to \eta + p \tag{5.1}$$

was calculated using a program developed at the University of Bonn[2]. For the beam energy different values between 800-1400 MeV were used in steps of 100 MeV. The Figures 5.1, 5.2 and 5.3 show the results of these calculations. In this respect an information of rather high importance is the fact, that the maximum θ-angle of the backscattered proton is $\approx 50°$. Hence, in the analysis of experimental data all events that contain protons with larger θ-angles can be dropped, without loosing real η-Dalitz events.

5.1.2. ω-Production Calculations

In the same manner a two body calculation of the ω-production off the proton in a photo-nuclear process was performed using different incident energies above the production threshold of the ω-meson.

$$\gamma_{\text{beam}} + p_{\text{target}} \to \omega + p \tag{5.2}$$

In Figure 5.4 the θ angle of the backscattered proton is plotted versus the θ angle of the produced ω-meson. As can be learned from the displayed result, the maximum θ-angles

---
[2]University of Bonn, http://pi.physik.uni-bonn.de/.

5.1. Two Body Calculations

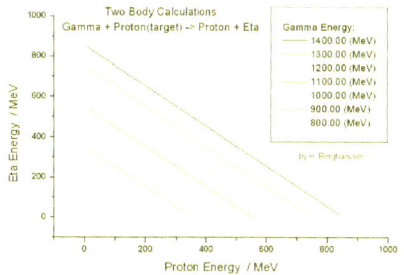

Figure 5.1.: Plot of the η-energy against the energy of the proton.

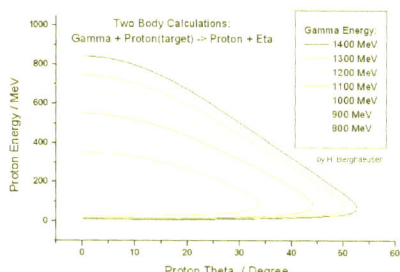

Figure 5.2.: The energy of the backscattered proton is plotted against the proton θ-angle.

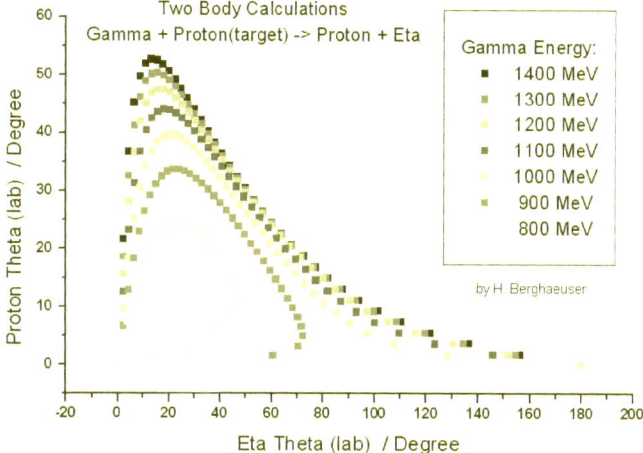

Figure 5.3.: Calculation for different incident energies: θ-angle of the proton plotted versus the θ-angle of the η-meson.

for the proton and the ω-meson are $33°$ and $40°$; these information were exploited in the analysis of experimental data in order to eliminate background events. In Figure A.10 (appendix) the proton energy is plotted versus the θ-angle of the proton.

5. Simulation

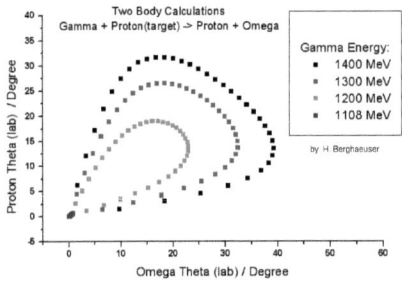

Figure 5.4.: Calculation for different incident energies: θ proton plotted versus θ of the ω-meson.

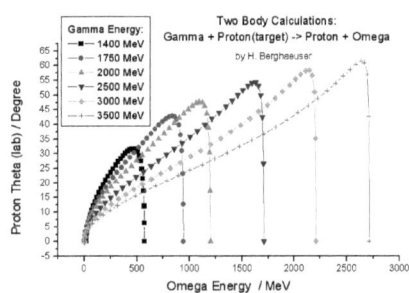

Figure 5.5.: Calculation for different incident energies: θ proton plotted versus energy of the ω-meson.

5.1.3. π^0-Production Calculations

The same calculations were performed for the π^0-meson. The implemented reaction is:

$$\gamma_{\text{beam}} + p_{\text{target}} \rightarrow \pi^0 + p \qquad (5.3)$$

This reaction was calculated for different incident energies between 200 and 1400 MeV. The results are shown in the Figures 5.6, 5.7, and 5.8. The obtained information can be used to define two dimensional cuts[3], which could be used to reduce the number of background events.

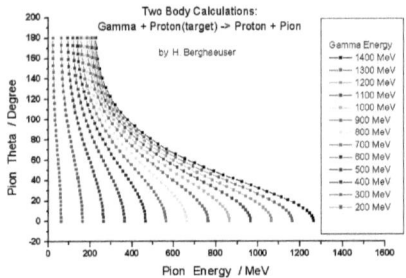

Figure 5.6.: Calculation for different incident energies: θ-angle of the produced π^0 plotted versus its energy.

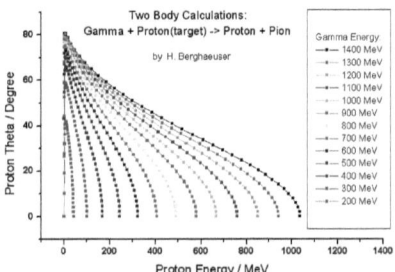

Figure 5.7.: Calculation for different incident energies: θ-angle of the backscattered proton plotted versus its energy.

---
[3]E.g. based on the ROOT-class TCutG.

5.2. Start Distributions of the Dalitz simulations

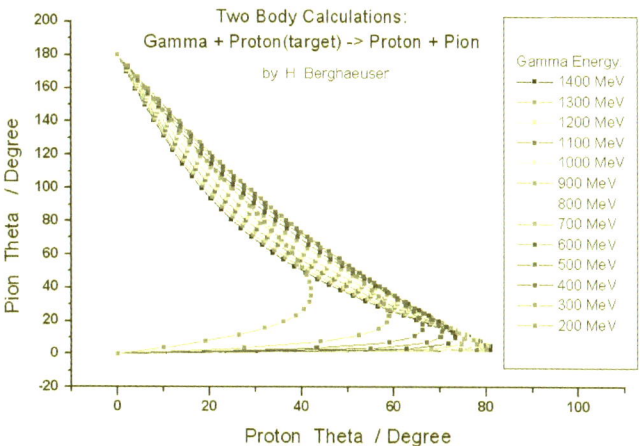

Figure 5.8.: Calculation for different incident energies: the plot shoes the θ-angle of the produced π^0 versus the θ-angle of the backscattered proton.

5.1.4. η'-Production Calculations

In a last *two body* calculation the kinematics of the production of η' were investigated. The following reaction was studied for an incident energy of 1450 MeV.

$$\gamma_{\text{beam}} + p_{\text{target}} \to \eta' + p \qquad (5.4)$$

The results are shown in the Figures 5.9 and 5.10.

5.2. Start Distributions of the Dalitz simulations

In order to apply an acceptance correction based on the detector response for a certain Dalitz decay, the invariant e^+e^--mass spectrum obtained in an analysis[4] of simulated data has to be divided by the corresponding spectrum of the start distribution. Thus it is important to use an event-generator capable of generating the correct distributions; otherwise it would not be possible to determine an exact detector acceptance. An event generator matching this requirement is the PLUTO event generator.
All Dalitz start distributions are discussed in the following subsections.

---

[4]This implies, that the same cuts are applied on the simulated data, which are applied in the corresponding analysis of the experimental data.

5. Simulation

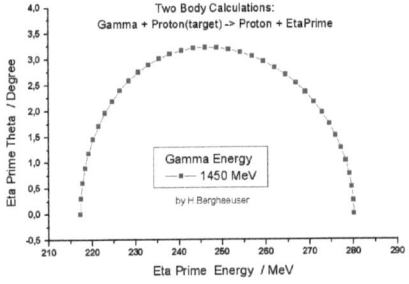
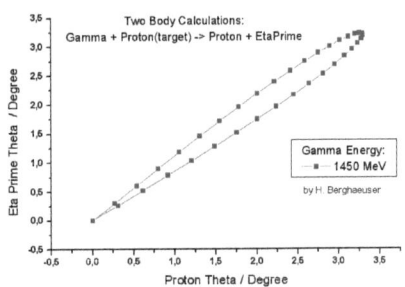

Figure 5.9.: θ-angle of the produced η' plotted versus its energy.

Figure 5.10.: θ-angle of η' plotted against the θ-angle of the backscattered proton.

5.2.1. η-Dalitz Phase Space Distribution

The event generator used to generate phase space distributions was described in section 4.1.4. As the PLUTO event generator was not usable in the beginning, because its output was not compatible to the A2 MC-simulation program, a preliminary Monte Carlo simulation was performed using a phase space distribution. Despite the fact, that the distribution of the invariant (e^+e^-)-mass in the start distribution was not realistic, the results of 500.000 simulated events delivered important information. Based on these a preliminary estimation of the detector response was worked out and moreover the graphical $dEversusE$ cuts for the PID and the VETO as well as the *time of flight* cuts could already be defined.

Figure 5.11 shows the $1/E$ distribution of the beam photons of the start distribution. The (incorrect) distribution of the invariant $mass_{e^+e^-}$ is shown in Figure 5.12. As soon as a start distributions generated by the PLUTO event generator were usable in connection with the standard A2 simulation program, new start distributions were generated for all Dalitz decays under investigation (see the following sub sections). However, one might wonder how strong the difference is between MC-simulations using a phase space and a PLUTO distribution. Concerning this question a simple comparison was worked out as soon as the first PLUTO event generation of 250.000 η-Dalitz events was accomplished. This comparison is described in subsection 5.2.5.

5.2.2. η-Dalitz PLUTO Distribution

The PLUTO event generator has already been described in section 4.1.4. It is capable of generating realistic Dalitz distributions. PLUTO was developed as an event generator for the simulations concerning the HADES@GSI experiment. As it turned out to be a

5.2. Start Distributions of the Dalitz simulations

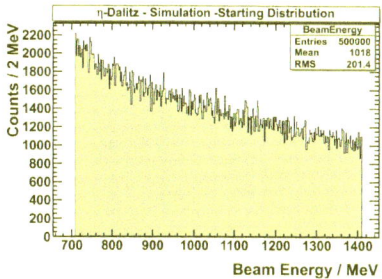
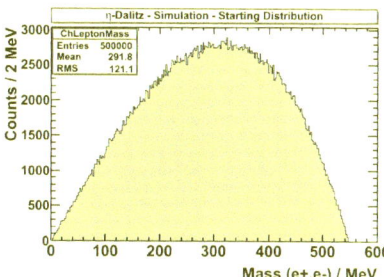

Figure 5.11.: Phase space start distribution: Energies of the γ-beam following a $1/E$-distribution.

Figure 5.12.: Phase space start distribution: invariant e^+e^--mass.

trustworthy and flexible generator the decision was made to use it for A2-simulations as well.

The problem with PLUTO was, that the GEANT4 based MC-simulation program of the A2-Collaboration (called 'A2sim') was not capable of decoding the data format of the output files that PLUTO produced. The solution was found by D. Werthmüller[5] who programmed a suitable converter that simply turns the output files of PLUTO into a certain data format which could be decoded by the *A2sim* program.

Using the PLUTO event generator 250.000 events of the following type were generated in a first step:

$$\gamma + p \rightarrow \eta + p \rightarrow \gamma + \gamma^* + p \rightarrow \gamma + e^- + e^+ + p \quad (5.5)$$

As it turned out, that more statistics were needed to determine an accurate detector response depending on the invariant mass of e^+e^-, more events were generated and simulated[6]. Figure 5.14 shows the invariant e^+e^--mass of five million generated events of the start distribution (using the beam energies shown in Figure 5.13). In the Figures 5.15 and 5.16 the distributions in θ and ϕ of the generated η-mesons are shown.

5.2.3. ω-Dalitz PLUTO Distribution

As with the generation of η-Dalitz events, the PLUTO generator was used to generate five million ω-Dalitz events. The implemented reaction is given by:

$$\gamma + p \rightarrow \omega + p \rightarrow \pi^0 + \gamma^* + p \rightarrow \pi^0 + e^- + e^+ + p \quad (5.6)$$

---

[5]D. Werthmüller, University of Basel, Dominik.Werthmueller@unibas.ch .

[6]In a second step 2.5 million, in a 3rd step 5 million events were generated and simulated.

5. Simulation

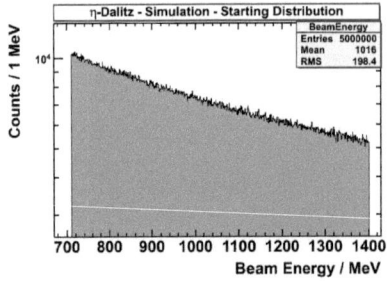

Figure 5.13.: Pluto start distribution: γ-beam flux following a $1/E_\gamma$ energy distribution.

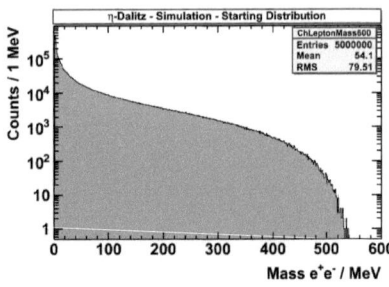

Figure 5.14.: Pluto start distribution: the invariant e^+e^--mass of the generated η-Dalitz decays.

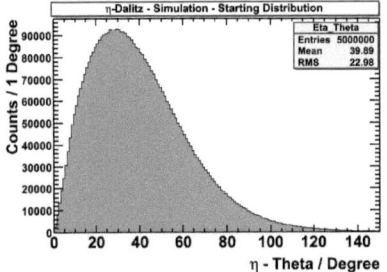

Figure 5.15.: Pluto start distribution: the distribution in θ of the generated η-mesons.

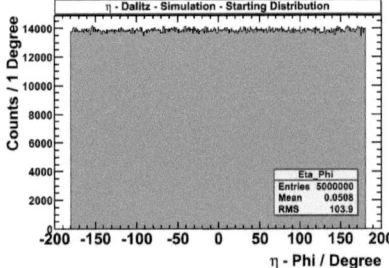

Figure 5.16.: Pluto start distribution: the distribution in ϕ of the generated η-mesons.

The Figure 5.17 shows the interval of incident γ-beam energies. In Figure 5.18 the resulting invariant e^+e^--mass of the start distribution is shown.

5.2.4. π^0-Dalitz PLUTO Distribution

The same procedure was accomplished in the case of the π^0-Dalitz decay. This time the following raction was implemented into PLUTO:

$$\gamma + p \rightarrow \pi^0 + p \rightarrow \gamma + \gamma^* + p \rightarrow \gamma + e^- + e^+ + p \tag{5.7}$$

Figure 5.19 shows the range of incident γ-beam energies. In Figure 5.20 the resulting invariant e^+e^--mass of the start distribution is shown.

5.2. Start Distributions of the Dalitz simulations

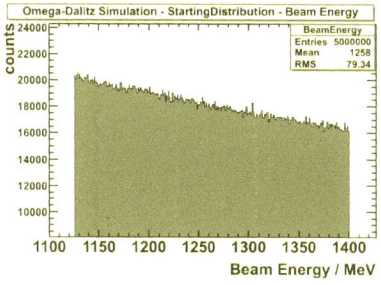

Figure 5.17.: Pluto start distribution: γ-beam flux following a $1/E_\gamma$ energy distribution.

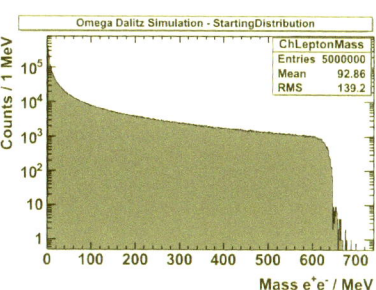

Figure 5.18.: Pluto start distribution: the invariant e^+e^--mass of the generated ω-Dalitz decays.

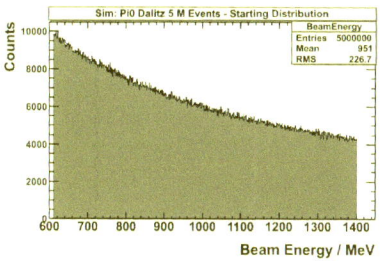

Figure 5.19.: Pluto start distribution: γ-beam flux following a $1/E_\gamma$ energy distribution.

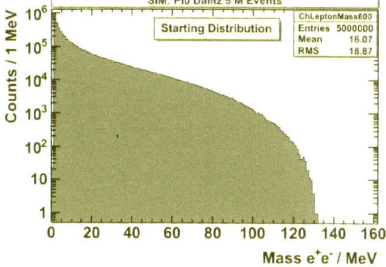

Figure 5.20.: Pluto start distribution: the invariant e^+e^--mass of the generated π^0-Dalitz decays.

5.2.5. Comparison between Start distributions: PLUTO vs. Phase Space

Concerning the generation of Dalitz events both event generators clearly have to produce a different output. The important question is how different the resulting acceptances are.
Although all other analyses of simulated data are presented and discussed in chapter 6, this special comparison shall be addressed in the following.

As a simulation of 0.5 million η-Dalitz events using the phase space distribution was already performed before the PLUTO event generator could be used, this comparison

5. Simulation

could be investigated rather fast once the PLUTO output could be used. This comparison was accomplished using the first performed Pluto simulation, in which 250.000-η-Dalitz events were investigated. It was correctly assumed that the differences mainly occur for the opening angle between e^+ and e^- and the distributions of the invariant (e^+e^-)-mass of the decay products of the η-meson. As other variables such as the missing mass, coplanarity, momentum balance etc. did not show any general difference in shape, the corresponding histograms will not be presented.

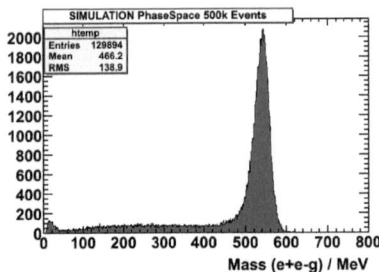 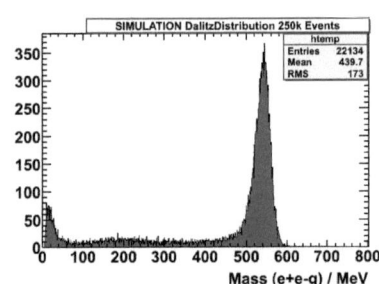

Figure 5.21.: Phase Space: invariant mass of $e^+e^-\gamma$.

Figure 5.22.: Pluto: invariant mass of $e^+e^-\gamma$.

While in the distribution of the invariant $e^+e^-\gamma$-masses both results differ only slightly (Figures 5.21 and 5.22), a strong and obvious difference can be seen in the two dimensional plots of $m_{e^+e^-}$ versus $m_{e^+e^-\gamma}$ (Figure 5.23 and Figure 5.24). This is due to the different distributions of the invariant masses of e^+e^- (Figure 5.25 and 5.26).

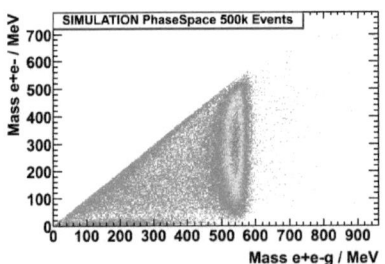 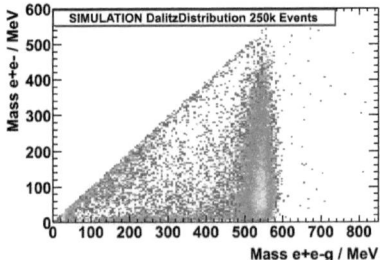

Figure 5.23.: Phase Space: invariant mass of e^+e^- plotted against $m_{e^+e^-\gamma}$.

Figure 5.24.: Pluto: invariant mass of e^+e^- plotted against $m_{e^+e^-\gamma}$.

5.2. Start Distributions of the Dalitz simulations

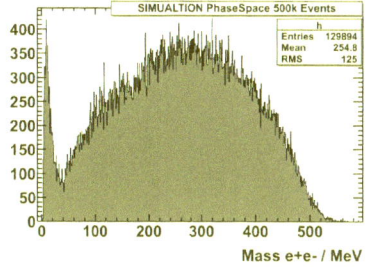

Figure 5.25.: Phase Space: invariant mass of $e^+e^-\gamma$.

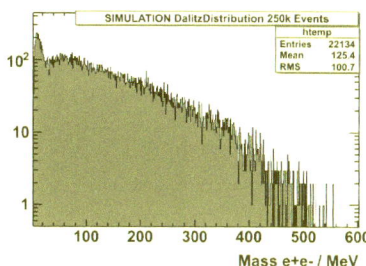

Figure 5.26.: Pluto: invariant mass of $e^+e^-\gamma$.

A distinct difference can be seen in the distributions of e^+e^--opening angles and is shown in Figures 5.27 and 5.28. As a consequence, the opening angles of (e^+, γ) and (e^-, γ) have to differ too (Figure 5.29 - 5.32).

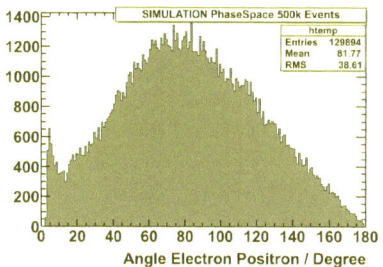

Figure 5.27.: Phase Space: reconstructed opening angle of (e^+, e^-).

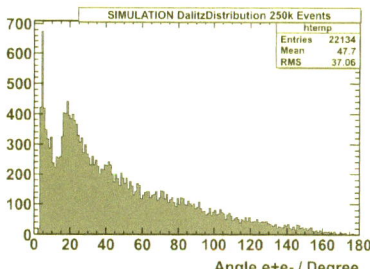

Figure 5.28.: Pluto: reconstructed opening angle of (e^+, e^-).

The θ-angles of the electrons and positrons show only very slight differences. In case of the events generated by PLTUO the distributions of θ are shifted a little bit towards smaller angles (TAPS region). As the CB/TAPS experiment does not use a magnetic field, the sign of a charged particle can not be determined. In an event, hits in the CB are processed before the hits in TAPS. Furthermore in a final state like $e^+e^-\gamma$, the charged lepton processed first is marked as e^-. Hence, e^- is more likely to be a CB-hit, whereas e^+ is more likely to be a TAPS-hits. This explains the differences between the angular distributions shown in Figures 5.33 and 5.35.

5. Simulation

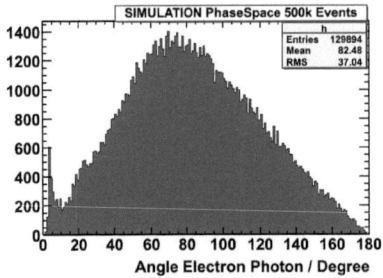

Figure 5.29.: Phase Space: reconstructed opening angle of (e^-, γ).

Figure 5.30.: Pluto: reconstructed opening angle of (e^-, γ).

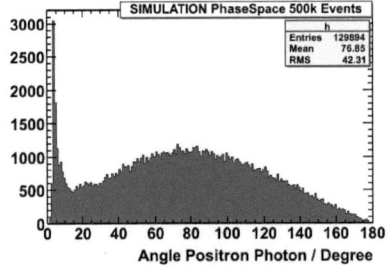 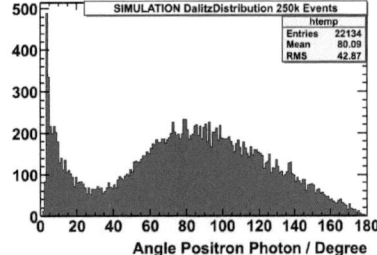

Figure 5.31.: Phase Space: reconstructed opening angle of (e^+, γ).

Figure 5.32.: Pluto: reconstructed opening angle of (e^+, γ).

As far as the θ-angles of the photon and the proton are concerned no difference can be detected (as assumed - Figures 5.35 - 5.40). The same holds for all ϕ-angle distributions (Figures 5.41 to 5.48).

However, the major difference is in the distribution of the invariant mass of e^+e^-, which directly affects the acceptance. Hence, the Pluto event generator delivers the more realistic results and for this reason it was used for all subsequent simulations of Dalitz decays.

5.2. Start Distributions of the Dalitz simulations

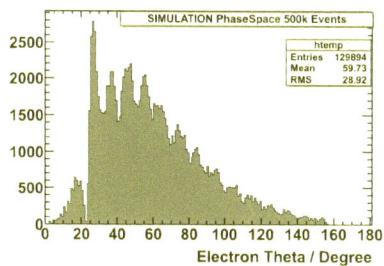

Figure 5.33.: Phase Space: reconstructed θ-angle of the electron.

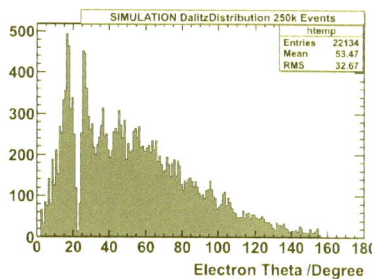

Figure 5.34.: Pluto: reconstructed θ-angle of the electron.

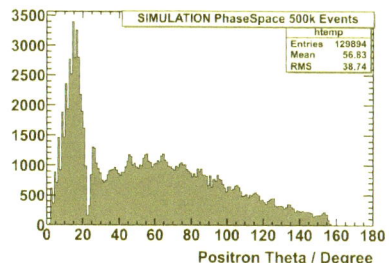

Figure 5.35.: Phase Space: reconstructed θ-angle of the positron.

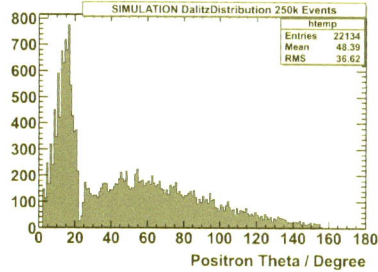

Figure 5.36.: Pluto: reconstructed θ-angle of the positron.

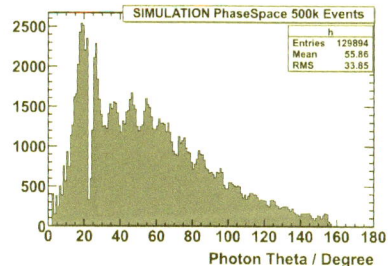

Figure 5.37.: Phase Space: reconstructed θ-angle of the photon.

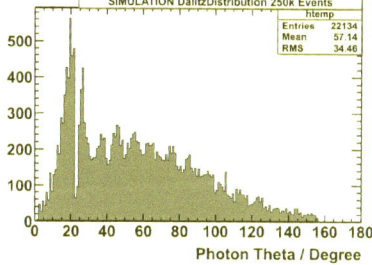

Figure 5.38.: Pluto: reconstructed θ-angle of the photon.

5. Simulation

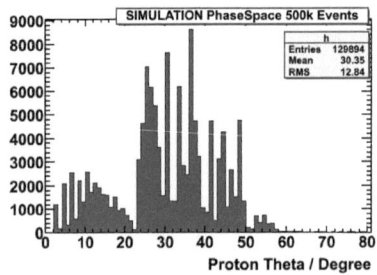

Figure 5.39.: Phase Space: reconstructed θ-angle of the proton. The reduced yield near $\theta = 20°$ is due to the gap between TAPS and CB.

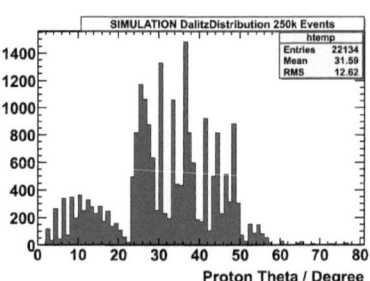

Figure 5.40.: Pluto: reconstructed θ-angle of the proton.

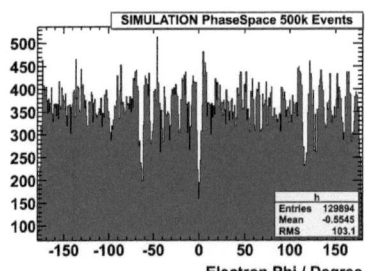

Figure 5.41.: Phase Space: reconstructed ϕ-angle of the electron. The reduced yield near $\phi = 0°$ is due to the CB support structure.

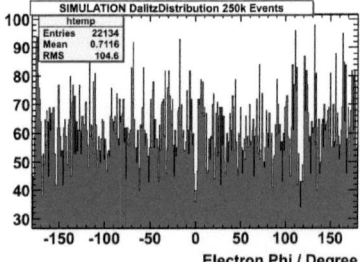

Figure 5.42.: Pluto: reconstructed ϕ-angle of the electron.

5.2. Start Distributions of the Dalitz simulations

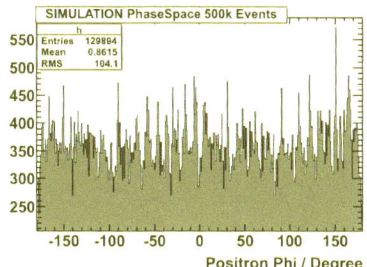

Figure 5.43.: Phase Space: reconstructed ϕ-angle of the positron.

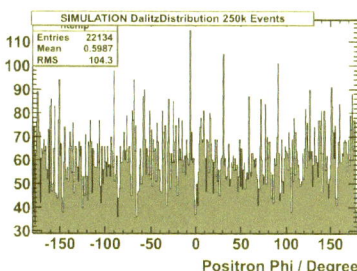

Figure 5.44.: Pluto: reconstructed ϕ-angle of the positron.

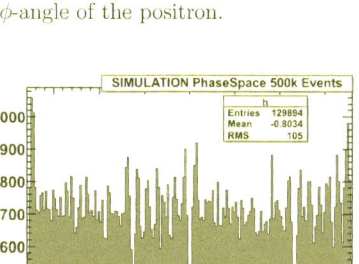

Figure 5.45.: Phase Space: reconstructed ϕ-angle of the photon.

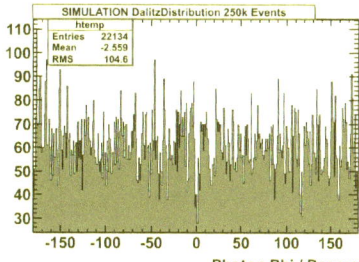

Figure 5.46.: Pluto: reconstructed ϕ-angle of the photon.

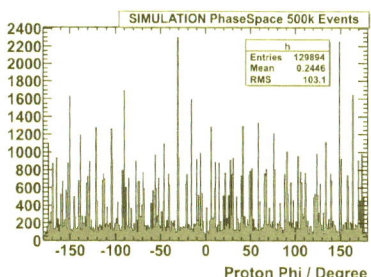

Figure 5.47.: Phase Space: reconstructed ϕ-angle of the proton.

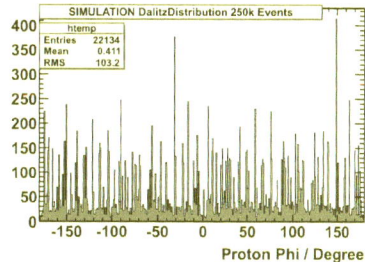

Figure 5.48.: Pluto: reconstructed ϕ-angle of the proton.

5. Simulation

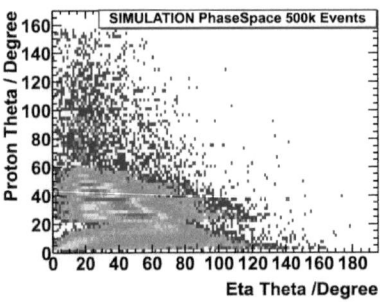

Figure 5.49.: Phase Space: 2D-Plot of θ-proton versus the θ-angle of the produced η-meson.

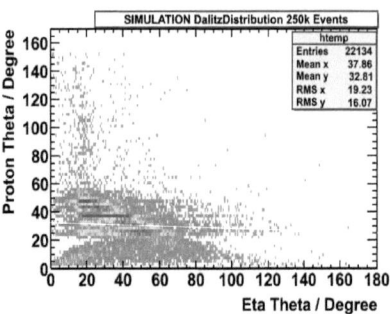

Figure 5.50.: Pluto: 2D-Plot of θ-proton versus the θ-angle of the produced η-meson.

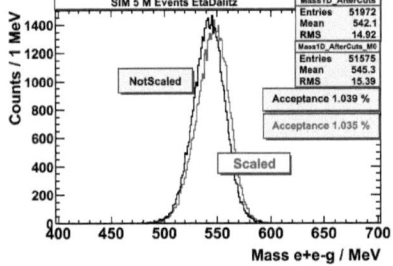

Figure 5.51.: The energy information of the CB and TAPS detectors in the Monte Carlo simulation with and without modified correction factor.

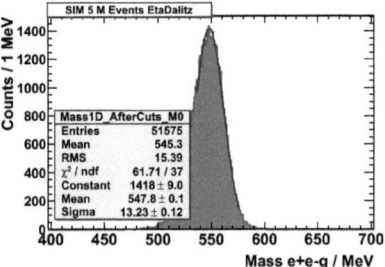

Figure 5.52.: MC-simulation: invariant $e^+e^-\gamma$-mass distribution after a proper scaling of the energy, the η-mass is located at 547.8 MeV (fit).

5.3. Energy scaling

It was found, that in the Monte Carlo simulations the reconstructed invariant masses of the η- and the ω-meson were slightly off. This problem was solved rather fast by increasing the *energy scaling factors*. The factor in the MC-configuration file of the Crystal Ball was increased from 1.07 to 1.082; in the configuration file of the BaF_2-calorimeter the corresponding factor was increased from 1.0 to 1.07.

Figure 5.51 shows the reconstructed[7] invariant η-masses of two 5 million simulated η-Dalitz events for both situations: with and without the enhancement of the energy scaling factor. In case of the former the reconstructed invariant η-mass is 547.8 MeV and corresponds to the PDG value (Figure 5.52). The width of the invariant mass distribution is $\sigma = 13$ MeV corresponding to an invariant mass resolution of 2.4 %.

---

[7]The principle steps in the process of reconstruction is explained in chapter 4. The applied cuts and more details on the analysis are described in the following chapter 6.

5. Simulation

6. Analysis

In this chapter the results of all accomplished analyses of the various decay channels of the π^0, η, ω-mesons are presented.

First the investigations of the simulated reactions and decay channels are discussed. Thereafter the analyses of the same decays in the experimental data are presented.
The analysis of the η-Dalitz decay is presented in the sections 6.1.3 and 6.2.4.
In all performed analyses several cuts on the kinematics (e.g. *missing mass, energy balance*, etc.) were applied. These cuts are listed in tables; in the tables the unit of the missing mass, energy balance, beam energy and the momentum is *MeV*. The unit for all angles is *degree*.

6.1. Analysis of simulated data

Monte Carlo simulations provide a very good mean to determine the detector acceptance for each decay channel. Furthermore characteristics of the decays as well as possible background channels can be investigated. Moreover the application of cuts can be tested and the loss of *real* events can be analyzed.
Once the acceptance of the detector system for a certain channel has been determined, the results of the analyses of the experimental data can be corrected; moreover the number of originally produced mesons in the data-set can be calculated[1].

In the analysis of simulated data (as well as in the investigation of the experimental data) a lot of different cut-settings have been tested. Presenting all tested cut-settings would definitely go beyond the scope of this thesis[2]; thus only the results of the final and most appropriated cut-settings will be presented and discussed. If one wants to perform an acceptance correction all applied cuts in the analysis of the simulated data have to be equal to the ones applied in the analysis of the experimental data; which was realized in all accomplished analyses. As an only exception the cuts on the prompt peaks in the timing spectra of the detected particles in the experimental data have not

---
[1] Using the corresponding branching ratio listed in the Particle Data Book published by the Particle Data Group.
[2] E.g. only in case of the η-Dalitz analysis 67 different cut-settings were tested.

6. Analysis

to be applied in the investigation of the simulated data, because in this case all 'hits' are always prompt (thus cuts on the prompt peaks are always fulfilled).

6.1.1. SIM: Exclusive analysis of $\eta \to \gamma\gamma$

A very important decay channel of the η-meson is $\eta \to \gamma\gamma$, which has a branching ratio of $(39.38\pm0.26)\%$ [20]. As this decay channel has a large branching ratio it can be easily measured. An analysis of $\eta \to \gamma\gamma$ in the data provides the possibility to determine the exact number of produced η-mesons in the experiment with a high accuracy. In order to determine this number the acceptance has to be determined first, which can be done by using a MC-simulation. As has been described in the previous chapter 10 million events were simulated and analyzed.

In the analysis of the simulated data-set the same *two-gamma-analysis-function* was used as for the analysis of the real data. In this function all events that contained at least two detected photons and one proton were analyzed; further on all events with higher multiplicities, that means more detected protons and/or photons, were analyzed too.

Higher multiplicities can easily be generated by so-called split-offs[3]. In case of the analysis of $\eta \to \gamma\gamma$ the higher multiplicities are handled in the following way: for each event all possible combinations of always two photons and one proton out of all available particles is built and analyzed. One important aspect when dealing with higher multiplicities is, to ensure, that no event is counted twice in the final results, which can easily be realized by checked the *event-numbers*[4].

The detector acceptance for events containing $\gamma\gamma$ and one proton is 47.5%. After application of cuts the determined acceptance is:

$$Acc_{\eta \to \gamma\gamma} = 12.1\% \quad \text{(p required)}$$

It has to mentioned again, that these cuts have to be equal to the cuts applied in case of the *experimental* data (except for cuts on the timing).

Table 6.1 lists all applied cuts in the exclusive analysis of simulated $\eta \to \gamma\gamma$ events.

Most cuts of this cut-setting (Table 6.1) correspond to the cuts applied in the search for

---

[3] *Split-off*: whenever a particle hits the calorimeter a cluster will be created. In case of protons (neutrons) these clusters mostly consists of only one or two crystals; but when photons or electrons hit the calorimeter an electromagnetic shower will be produced that spreads over several crystals (up to 15). It can happen, that the produced e^+/e^- of a shower do not fire a single crystal they are passing through but the next crystal they enter. In this case the cluster routine detects a second 'hit', which in reality is only a split off of the the first hit.

[4] Each event has its own unique number (ID).

η-Dalitz decays. The reason for this is to have a better comparison of the results (and number of reconstructed events - in the *experimental* data).

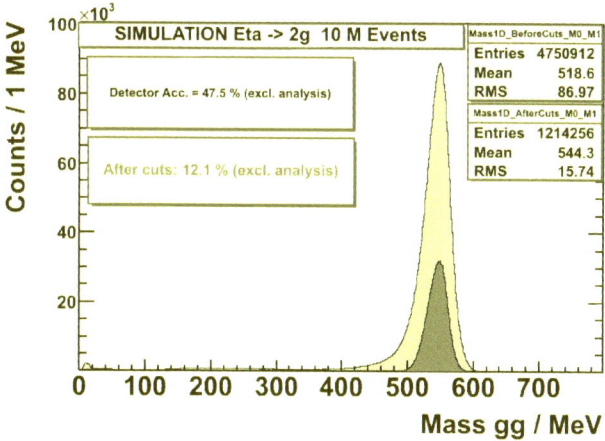

Figure 6.1.: MC-Simulation of 10.000.000 events of $\eta \to \gamma\gamma$.

One appropriate mean to suppress events containing split-offs is to raise the cluster threshold in the configuration files of the calorimeters (NaI, BaF_2). This is important as split-offs always lead to a loss of shower-energy, which shall be explained by an easy example.
Assume $\omega \to \pi^0 \gamma$ shall be reconstructed. When the γ produces a split-off in a calorime-

Cut	Min	Max
BeamEnergy	750.0	1210.0
ProtonTheta	0.0	50.0
Momentum-X	-40.0	40.0
Momentum-Y	-40.0	40.0
Momentum-Z	-100.0	105.0
Missing Mass	910.0	975.0
Coplanarity	168.0	192.0
	-168.0	-192.0

Table 6.1.: Applied cuts.

6. Analysis

ter, the final state will be equal to $4\gamma^5$. The analysis routines will generate all possible combinations picking 3 photons out of the 4 available ones. As a result, the calculated Lorentz vector of the ω meson is not correct, because energy is missing (energy of the split-off hit). Thus it is important to remove split-off events.

The Figures 6.2, 6.3, 6.4 and 6.5 show the results of an exclusive and an inclusive analysis of $\eta \to \gamma\gamma$ for two different cluster thresholds (20 MeV and 50 MeV). For all performed analyses in this work a cluster threshold of 50 MeV was used.

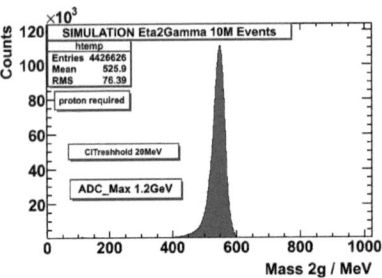

Figure 6.2.: Simulation of $\eta \to \gamma\gamma$ (detection of proton required). The cluster thresholds of CB and TAPS was set to 20 MeV.

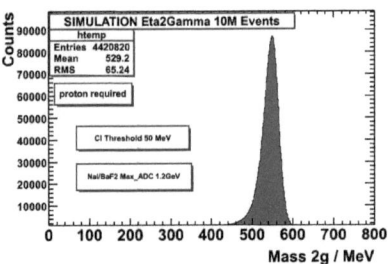

Figure 6.3.: Simulation of $\eta \to \gamma\gamma$ (detection of proton required). The cluster thresholds of CB and TAPS was set to 50 MeV.

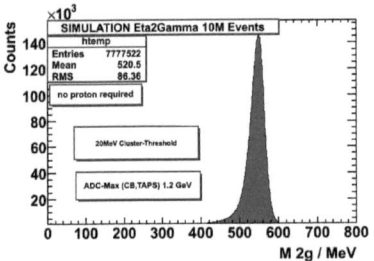

Figure 6.4.: Simulation of $\eta \to \gamma\gamma$ (no proton required). The cluster thresholds of CB and TAPS was set to 20 MeV.

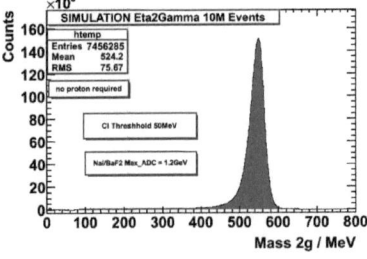

Figure 6.5.: Simulation of $\eta \to \gamma\gamma$ (no proton required). The cluster thresholds of CB and TAPS was set to 50 MeV.

---

[5]With $\pi^0 \to \gamma\gamma$.

6.1. Analysis of simulated data

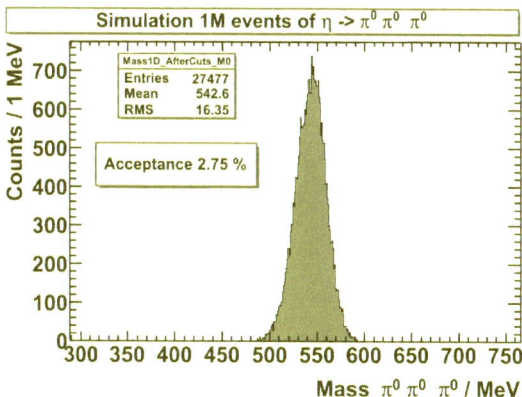

Figure 6.6.: Simulation: reconstructed invariant mass spectrum of $3\pi^0$ after cuts.

6.1.2. SIM: Exclusive analysis of $\eta \to \pi^0\pi^0\pi^0$

The decay of the η-meson into three π^0-mesons is the second strongest η-decay and has a branching ratio of 32.31 ± 0.23%. In the analysis of the final sate of '6γ proton' higher multiplicities were not investigated (neither in the case of simulated data nor in the case of experimental data). The cuts, that were applied, are listed in Tabel 6.2; in this Table *Coplanarity* stands for the absolue value.

Cut	Min	Max
BeamEnergy	750.0	1210.0
ProtonTheta	0.0	50.0
Momentum-X	-50.0	50.0
Momentum-Y	-50.0	50.0
Momentum-Z	-100.0	105.0
Missing Mass	910.0	975.0
Energy balance	-50.0	50.0
Coplanarity	168.0	192.0

Table 6.2.: Applied cuts.

Of the one million events, that were started in MC-simulation, only 27477 events were reconstructed and survived the cuts. Thus the corresponding acceptance is:

$$Acc_{\eta \to 3\pi^0} = 2.75\% \quad \text{(p required)} \tag{6.1}$$

The spectrum of the reconstructed invariant $3\pi^0$-mass is shown in Figure 6.6.

6.1.3. SIM: Exclusive analysis of $\eta \to e^+e^-\gamma$

Five million events of $\eta \to e^+e^-\gamma$ were simulated and analyzed exclusively. Again the same analysis-function was used as for the corresponding investigation of the experimen-

6. Analysis

tal data.

All events containing at least one photon, one proton and two additional charged hits (electrons) were investigated. Again, higher multiplicities were taken into account. In this respect all events with any greater number of protons and photons and up to seven additional charged hits were analyzed; whereas for each of these events all possible combinations leading to $p\,e^+e^-\gamma$ were generated[6].

In total 67 different cut-settings were tested both on real data and on the output of the MC-simulation. In case of the simulated data no cuts on the timing spectra were applied. The cut-setting found to be most appropriate is listed in Table 6.3. For this set of cuts the over all acceptance was determined as[7]:

$$Acc_{\eta \to e^+e^-\gamma} = 1.3\% \quad \text{(proton required)} \tag{6.2}$$

The reason for the different cuts on the cluster sizes of the positron and the electron

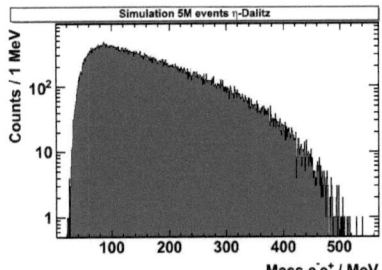 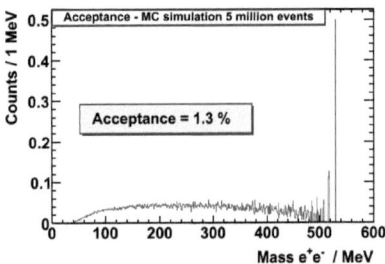

Figure 6.7.: Distribution of the detected e^+e^--mass in the η-Dalitz analysis of simulated data.

Figure 6.8.: Determined acceptance of $\eta \to e^+e^-\gamma$.

(see Table 6.3) is easy to be explained. As has been mentioned before it is not possible in the CB/TAPS experiment to distinguish between positive and negative charges[8]. Hence, the naming of e^+ and e^- is merely a formal aspect. But, in the reconstruction of the particles the cluster-routines of TAPS and Crystal Ball will always process the hits according to their energies (first the one with the highest energy deposit). In the analysis functions e^- always corresponds to the first charged hit (that is not a proton); thus on average the cluster size of the first hit is larger than for those of the later hits.

---

[6] And of course, as mentioned before, it was ensured that no event was counted twice (or even more often) in the final results.
[7] The acceptance has been determined for many different cut-settings, but only the actual used one will be presented.
[8] Because the experiment does not use a magnetic field.

6.1. Analysis of simulated data

The reason for applying cuts on the cluster size anyway has been discussed in chapter 4 already. These cuts help to reduce background stemming from charged pions, as these always create smaller clusters of responding crystals in the calorimeters.

The relative strength of each cut was tested in a simulation of 2.5 million η-Dalitz events. The simulated data was analyzed 13 times applying only a single cut each time. The result of this investigation is shown in Table A.1 (appendix).

As far as the simulated data of $\eta \to e^+e^-\gamma$ is concerned, these cuts might not be of real importance, as there is no charged pion background to be reduced. Nevertheless, because these cuts were applied on the real data, they had to be applied on the simulated data too in order to determine a realistic and accurate detector response (acceptance).

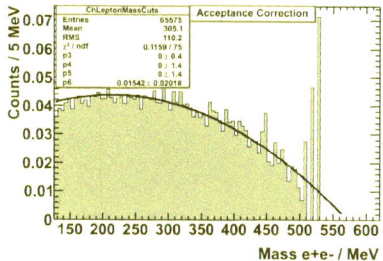

Figure 6.9.: Fit to the acceptance histogram.

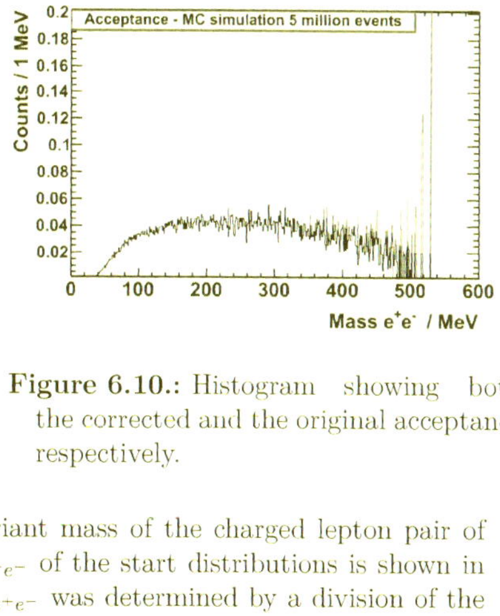

Figure 6.10.: Histogram showing both, the corrected and the original acceptance, respectively.

Figure 6.7 shows the distribution of the invariant mass of the charged lepton pair of reconstructed events; whereas the original $m_{e^+e^-}$ of the start distributions is shown in Figure 5.14. The acceptance depending on $m_{e^+e^-}$ was determined by a division of the distribution shown in Figure 6.7 by the original distribution (Figure 5.14). Figure 6.8 illustrates the resulting acceptance. Due to the low statistics in the regime of high invariant e^+e^--masses and the fact that the resolution effect is not present in the starting distribution, the high mass entries in the histogram of the acceptance show a *jagged* shape[9]. This jagged shape was corrected as follows: first the distribution was fitted[10]. The result is shown in Figure 6.9. Thereafter the jagged entries were corrected corresponding to the mean value of the fit in the corresponding mass regime. In Figure 6.10 both, the original and the corrected acceptance are shown. The Figures 6.11 and 6.12

---
[9]This jagged shape is generated by the division process.
[10]This was done for several acceptance histograms corresponding to the 67 different cut-settings, that were tested

6. Analysis

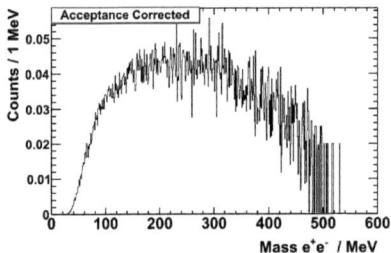

Figure 6.11.: The corrected acceptance used in the η-Dalitz analysis.

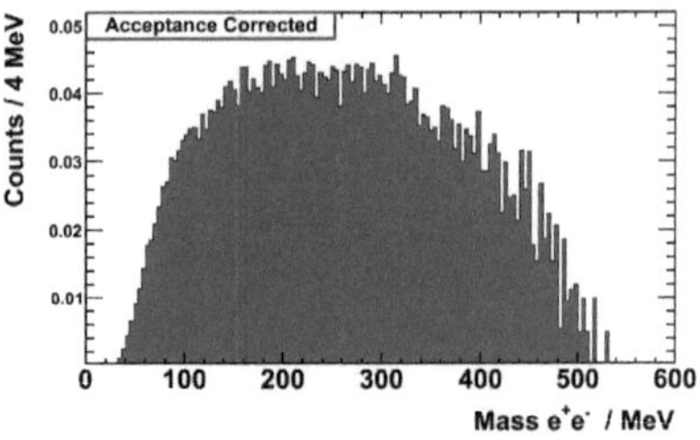

Figure 6.12.: Corrected acceptance with a binning of 4 MeV.

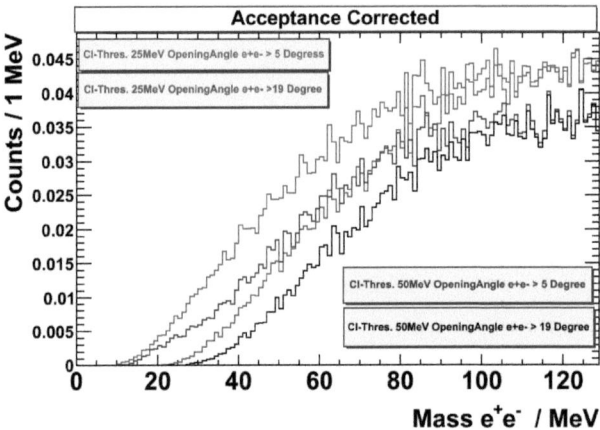

Figure 6.13.: Investigation of the shape of the acceptance for different cluster threshold and different cuts on the opening angle of e^+e^-.

show the corrected acceptance for different binning.

Furthermore the dependence of the detector response on certain cluster thresholds (25 MeV, 50 MeV) as well as cuts on the opening angle of e^+e^- was tested, which is shown in Figure 6.13. This investigation showed, that the acceptance for low invariant masses of e^+e^- is the larger the lower the cluster-threshold. Less strict cuts on the opening angle of e^+e^- increase the acceptance in the lower mass regime.

6.1. Analysis of simulated data

Cut	Min	Max
BeamEnergy	750.0	1210.0
ProtonTheta	0.0	50.0
Momentum-X	-40.0	40.0
Momentum-Y	-40.0	40.0
Momentum-Z	-100.0	105.0
Missing Mass	910.0	975.0
Energy Balance	-40	40
Coplanarity	168.0	192.0
Angle e^+e^-	19	140
Angle $e^+\gamma$	50	175
Angle $e^-\gamma$	50	175
e^- Cluster Size	5	14
e^+ Cluster Size	3	12

Table 6.3.: Applied cuts in the analysis of simulated and real-data events of $\eta \to e^+e^-\gamma$ (proton).

Cut	Min	Max
BeamEnergy	750.0	1210.0
ProtonTheta	0.0	50.0
Momentum-X	-40.0	40.0
Momentum-Y	-40.0	40.0
Momentum-Z	-100.0	105.0
Missing Mass	910.0	975.0
Energy Balance	-40	40
Coplanarity	168.0	192.0
Angle $\pi^+\pi^-$	9	180

Table 6.4.: Applied cuts in the analysis of simulated events $\eta \to \pi^+\pi^-\gamma$ (proton).

In order to fully suppress background stemming from conversion processes[11] a minimum angle of 19° between e^+e^- was required (section 6.4.1).
The reason for using the higher cluster threshold of 50 MeV in the analyses is simple to be explained. When using lower thresholds, the number of split-off effects will rise (e.g. in case of a 25 MeV the rise factor is 5.38). The η-Dalitz MC-simulation was performed for both cluster thresholds (20 MeV and 50 MeV). As in the authors investigations also an analysis of the ω-Dalitz was performed, an appropriate function was available for analyzing events of the type $e^+e^-\gamma\gamma$. This made a study of split-off effects in the case of the η-Dalitz analysis (of simulated events) possible. Figure 6.14 shows the result of this investigation. When using a cluster threshold of 50 MeV the probability for a split-off effect is less than 0.14%, thus it is a per mill effect.

6.1.4. SIM: Exclusive analysis of $\eta \to \pi^+\pi^-\gamma$

The analysis of $\eta \to \pi^+\pi^-\gamma$ is difficult, due to the channel $\eta \to \pi^0\pi^+\pi^-$. The latter has a larger branching ratio ($\approx 22.73\%$) than the former ($\approx 4.6\%$). If one of the photons stemming from the pion[12] is not detected, this will always lead to the same final state $\pi^+\pi^-\gamma$. Hence, this fact leads to a strong contribution to the background, which makes an analysis more difficult. One can apply cuts on the kinematics (missing mass, momentum balance, etc.), but as these cuts have a certain widths, a large fraction of

---
[11] A conversion process: $\gamma \to e^+e^-$
[12] $\pi^0 \to \gamma\gamma$, $BR = (98.798 \pm 0.032)\%$.

6. Analysis

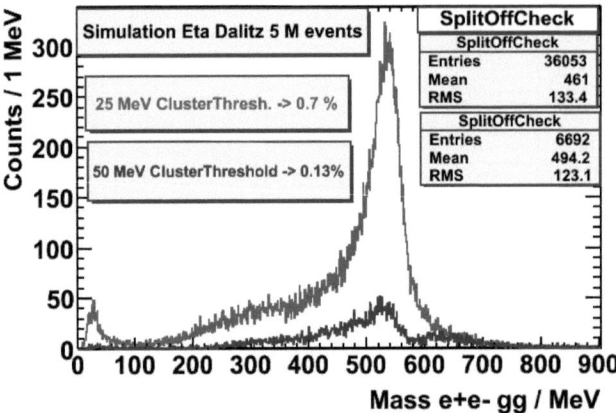

Figure 6.14.: Investigation of the split-off-effect. Five million events of $\eta \to e^+e^-\gamma$ (proton) were simulated and thereafter analyzed using a function only investigating the final state $e^+e^-\gamma\gamma$.

events stemming from $\eta \to \pi^0\pi^+\pi^-$ will still survive.

In order to perform an estimation of the contribution of this additional background 900.000 simulated events of $\eta \to \pi^+\pi^-\pi^0$ were investigated using an analysis-function for the final state of $\pi^+\pi^-\gamma$. The cuts that were applied are listed in Tabel 6.4. The result is shown in Figure 6.16. As one can obtain from this figure 955 counts survived the applied cuts and thus the propability for a $\pi^+\pi^-\pi^0$-event to enter in analysis of $\pi^+\pi^-\gamma$ (after cuts) is 0.106%. Still, this leads to an enormous contribution. Based on the branching ratio of $\approx 22.7\%$ the number of start events corresponds to ≈ 4 million η-mesons in total. With the branching ratio of $\pi^+\pi^-\gamma$, which is $\approx 4.6\%$, we can now calculate the number of η-mesons that decay into $\pi^+\pi^-\gamma$. This number is ≈ 182.000. After an analysis of 900.000 simulated events of $\pi^+\pi^-\gamma$, the acceptance was determined as 0.68% (Figure 6.15). Thus the estimated number of counts in an analysis of experimental data is 1240[13]. Consequently it is very difficult to analyse this channel, as the contribution of $\pi^+\pi^-\pi^0$ to the detected η-signal is of equal size (955 counts).

However, the author decided that this analysis can not be successfully accomplished in parallel to the difficult Dalitz analyses. Thus the decision was made to investigate $\eta \to \pi^0\pi^+\pi^-$ instead of $\eta \to \pi^+\pi^-\gamma$.

---

[13]Based on the assumption that ≈ 4 million η-mesons were produced.

6.1. Analysis of simulated data

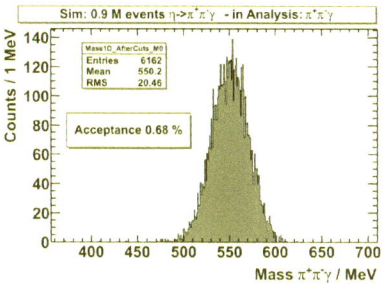

Figure 6.15.: Simulation of 0.9M events of $\eta \to \pi^+\pi^-\gamma$: spectrum of the reconstructed invariant mass.

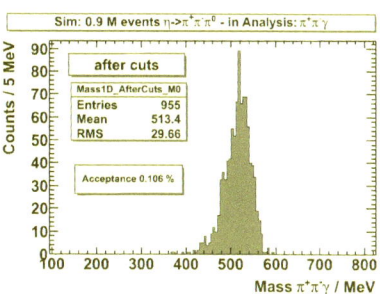

Figure 6.16.: Simulation: background in the analysis of $\pi^+\pi^-\gamma$ stemming from $\pi^+\pi^-\pi^0$.

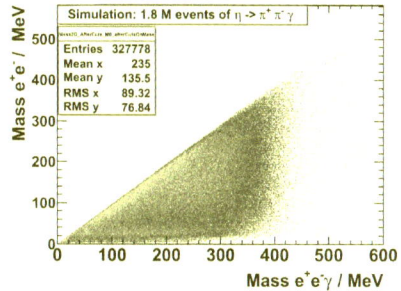

Figure 6.17.: Simulation of $\eta \to \pi^+\pi^-\gamma$. In total 18% enter in the η-Dalitz analysis (before cuts).

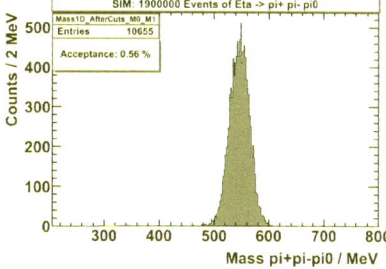

Figure 6.18.: Simulation: Invariant mass spectrum of $\eta \to \pi^+\pi^-\pi^0$ (proton) after cuts.

Nonetheless a second MC-simulation of 1.8 million events of $\eta \to \pi^+\pi^-\gamma$ was performed in order to investigate the contribution of this decay to the background in the η-Dalitz analysis. It has been found, that after the application of the cuts listed in Table 6.3 no events survive (Figure 4.12 and 4.13). Thus, this background channel can be successfully eliminated in the analysis of the η-Dalitz decay. Figure 6.17 shows the two dimensional plot of the invariant mass of e^+e^- versus the $e^+e^-\gamma$-mass; in total 18% of the original 1.8 million events are registered as $e^+e^-\gamma$ channel (including the backscattered proton).

6. Analysis

6.1.5. SIM: Exclusive analysis of $\eta \to \pi^+\pi^-\pi^0$

One of the aims was to measure the branching ratios of at least four η-decays. The reason for this was on the one hand to determine the number of produced η-mesons more then once (to be save) and on the other hand to check the outcome of the applied 'analysis-procedures' against published values in order to verify the results as well as the applied analysis-techniques. For this purpose the following four η-decays were chosen:

$$\eta \to \gamma\gamma$$
$$\eta \to \pi^0\pi^0\pi^0$$
$$\eta \to \pi^+\pi^-\pi^0$$
$$\eta \to e^+e^-\gamma$$

Hence, the detection efficiencies for all four channels had to be determined. As the corresponding results for $\eta \to \gamma\gamma$ and the η-Dalitz decay have already been presented in the previous two subsections, only the acceptance of the channels $\eta \to \pi^+\pi^-\pi^0$ and $\eta \to \pi^0\pi^0\pi^0$ remain to be presented.

Cut	Min	Max
BeamEnergy	750.0	1210.0
ProtonTheta	0.0	50.0
Momentum-X	-50.0	50.0
Momentum-Y	-50.0	50.0
Momentum-Z	-100.0	105.0
Missing Mass	915.0	985.0
Coplanarity	168.0	192.0
π^0 mass	120	150
π^- Cluster Size	0	7
π^+ Cluster Size	0	7

Table 6.5.: Applied cuts in the exclusive analysis of $\eta \to \pi^+\pi^-\pi^0$ (simulation).

Concerning the decay $\eta \to \pi^+\pi^-\pi^0$ 1.9 million events were simulated in a Monte Carlo simulation using a phase space start distribution. Thereafter the output of the Monte Carlo was analyzed using $AR_{HB}2v3$ under application of the same cuts, that were applied in the analysis of this decay in the experimental data. The cuts are listed in Table 6.5. Figure 6.18 shows the final histogram of the invariant mass distribution (after cuts). Again the detection of a proton was required and no cuts on the timing were applied. Using the listed cuts the detector acceptance for this η-decay channel is:

$$Acc_{\eta \to \pi^+\pi^-\pi^0} = 0.56\% \quad \text{(p required)} \tag{6.3}$$

6.1.6. SIM: Exclusive analysis of $\omega \to \pi^0\gamma$

One focus in the investigations of the A2 group of the University of Giessen is set on the study of in-medium effects of the ω-meson. In this context ω-production runs on the targets Nb and C were performed. The decay of interest is $\omega \to \pi^0\gamma$. Besides of the investigation of the Nb and C target runs the liquid Hydrogen runs were investigated. It is important to study ω-mesons that were produced on a proton target, as this provides the possibility to compare the measured ω-line shapes to a reference signal.

In the analysis of $\omega \to \pi^0\gamma$ different cut-settings were tested and the detector acceptance was determined for each setting. Not only events containing exactly three photons and one proton, but all events with a higher multiplicity (of photons and protons) were analyzed. A χ^2-test was used to identify the *best* pion[14] out of the three detected photons[15]. Figure 6.19 shows the reconstructed invariant $\pi^0\gamma$-mass in the exclusive analysis of two million simulated events. As 498369 simulated events are detected, the pure acceptance (without any cuts) is:

$$Acc_{\pi^0\gamma} = 24.9\% \quad \text{(p required - no cuts)} \tag{6.4}$$

After a strict cut on the π^0-mass (128-143 MeV) only 114889 events survive, which leads

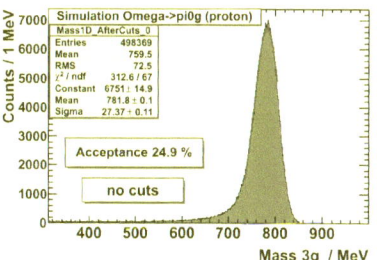

Figure 6.19.: Simulation: reconstructed invariant $\pi^0\gamma$-mass (no cuts).

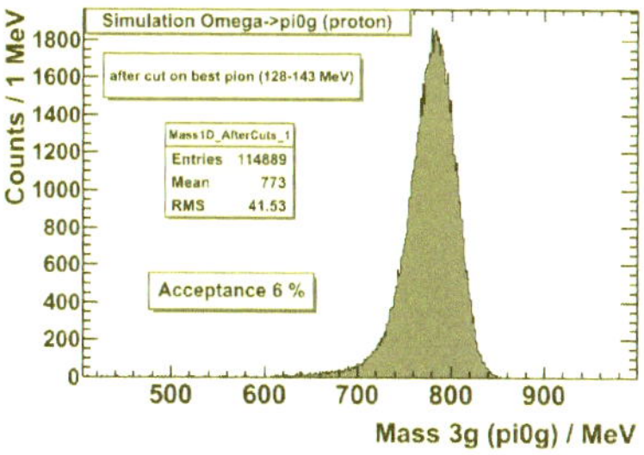

Figure 6.20.: Simulation: reconstructed invariant $\pi^0\gamma$-mass after a strict cut on the π^0-mass.

to an acceptance of 6% (Figure 6.20). The Figures 6.21 and 6.22 show the results for additional cuts on the θ-angle of the detected proton[16] and the coplanarity (165°-195°).

---

[14] $\pi^0 \to \gamma\gamma$

[15] This means only, that the best π^0 candidate was identified; no cut on the π^0-mass was applied during this step of the analysis.

[16] As was learned from a *two body simulation* the maximum θ-angle of a proton is 33° for ω-production off the proton and an incident beam erngy of 1.4 GeV.

6. Analysis

In the former case the acceptance is reduced to 4.1% and for the latter the acceptance is further reduced to 2.9%.

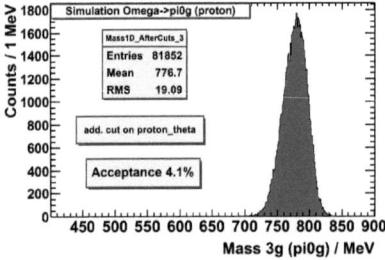

Figure 6.21.: Simulation: reconstructed invariant $\pi^0\gamma$-mass after an additional cut on the θ-angle of the detected proton.

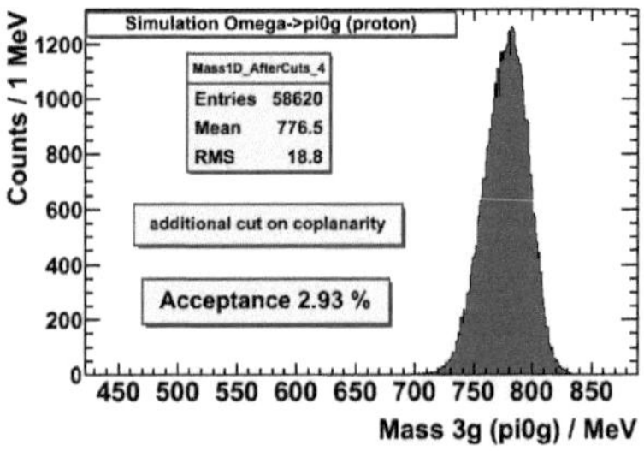

Figure 6.22.: Simulation: reconstructed invariant $\pi^0\gamma$-mass after an additional cut on the coplanarity.

Cut	Min	Max
BeamEnergy	1125.0	1400.0
ProtonTheta	0.0	33.0
Momentum-X	-20.0	20.0
Momentum-Y	-20.0	20.0
Momentum-Z	-40.0	90.0
Missing Mass	930.0	950.0
Coplanarity	170.0	190.0
Energy balance	-20	20
Pion Mass	128	143

Table 6.6.: Very strict 'Cut-Setting A'.

Cut	Min	Max
BeamEnergy	1125.0	1400.0
ProtonTheta	0.0	33.0
Momentum-X	-40.0	40.0
Momentum-Y	-40.0	40.0
Momentum-Z	-50.0	100.0
Missing Mass	915.0	965.0
Coplanarity	170.0	190.0
Energy balance	-40	40
Pion Mass	128	143

Table 6.7.: 'Cut-Setting B'.

The detection of the backscattered proton, and thus of all particles in the final state, provides every information required to apply additional cuts on the momentum balance and the energy balance. In the exclusive analysis of $\omega \to \pi^0\gamma$ different approaches were used. In oder to compare the results on lH_2 to the heavy target beamtimes (Nb, C), which were analyzed by M. Thiel[17], similar cuts[18] had to by applied.

Another aim was, to work out a very clean ω-signal; in other words to find a setting of

---
[17]Michaela Thiel, University of Giessen, Michaela.Thiel@exp2.physik.uni-giessen.de
[18]In this case: only cuts on the π^0-mass and the timing spectra were applied; whereas the latter are not applied in case of simulated data.

6.1. Analysis of simulated data

cuts resulting in the best possible *signal to background* ratio. Corresponding to this the Tables 6.6, 6.7 and 6.8 list different cut-settings that were tested. The Figures 6.23, 6.24 and 6.5 show the corresponding invariant mass spectra after cuts.

The determined acceptances for the specified cuts-settings A,B,C are:

A: $Acc_{\omega \to \pi^0 \gamma} = 1.108\%$

B: $Acc_{\omega \to \pi^0 \gamma} = 0.43\%$

C: $Acc_{\omega \to \pi^0 \gamma} = 1.87\%$

Cut	Min	Max
BeamEnergy	1125.0	1400.0
ProtonTheta	0.0	33.0
Momentum-X	-30.0	30.0
Momentum-Y	-30.0	30.0
Momentum-Z	-50.0	100.0
Missing Mass	925.0	955.0
Coplanarity	170.0	190.0
Energy balance	-30	30
Pion Mass	128	143

Table 6.8.: Very strict 'Cut-Setting C'.

Cut	Min	Max
BeamEnergy	1125.0	1410.0
ProtonTheta	0.0	33.0
Momentum-X	-50.0	50.0
Momentum-Y	-50.0	50.0
Momentum-Z	-100.0	105.0
Missing Mass	915.0	985.0
Coplanarity	168.0	192.0
π^0 mass	120	150
π^- Cluster Size	0	7
π^+ Cluster Size	0	7

Table 6.9.: Applied cuts in the exclusive analysis of $\omega \to \pi^+\pi^-\pi^0$ (simulation).

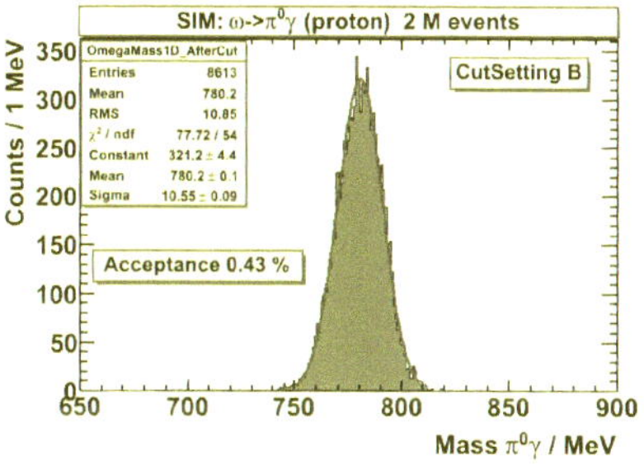

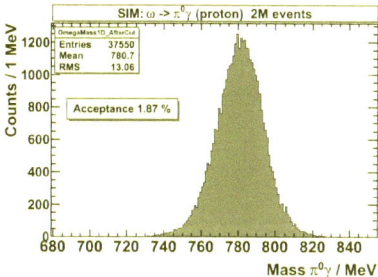

Figure 6.23.: Simulation: reconstructed invariant $\pi^0\gamma$-mass after application of cut-setting B.

Figure 6.24.: Simulation: reconstructed invariant $\pi^0\gamma$-mass after application of cut-setting C.

6. Analysis

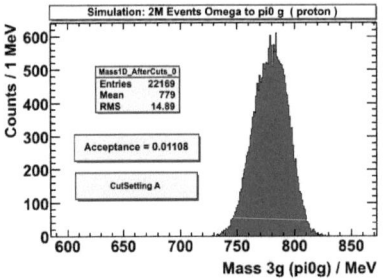
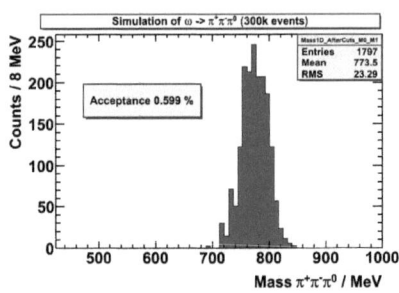

Figure 6.25.: Simulation: reconstructed invariant $\pi^0\gamma$-mass after application of cut-setting A.

Figure 6.26.: Simulation of $\omega \to \pi^0\pi^+\pi^-$: reconstructed ω-mass after cuts.

6.1.7. SIM: Exclusive analysis of $\omega \to \pi^+\pi^-\pi^0$

Besides the currently discussed decay of ω-meson two charged decay modes were investigated. These were the ω-Dalitz decay, which will be discussed in the following section, and the decay $\omega \to \pi^+\pi^-\pi^0$. This decay has the largest branching ratio $(BR = 89.1 \pm 0.7)\%$ of all ω-decays. Thus an analysis of this channel should provide results with an adequate statistics in order to determine the total number of produced ω-mesons in the data. Hence, the acceptance for this channel had to be determined.

As the final states of the channel of interest is equal to the final state of $\eta \to \pi^0\pi^+\pi^-$ (including the proton) the same analysis functions were used in $AR_{HB}2v3$[19]. All that was left to be done in order to determine the detector response was to perform an Monte Carlo simulation. The start distribution of 0.3 million events was generated using the *phase space event generator* (section 4.1.4). The output of the simulation was analyzed using the cuts listed in Table 6.8. Figure 6.26 shows the result. The total acceptance of this channel after application of the listed cuts is:

$$Acc_{\omega \to \pi^0\gamma} = 0.599\% \approx 0.6\% \qquad (6.5)$$

6.1.8. SIM: Exclusive analysis of $\omega \to \pi^0 e^+ e^-$

Five million events of $\omega \to e^+e^-\pi^0$ were simulated. As has already been described in chapter 5 the PLUTO event generator was used to generate the start distribution. The output of the MC-simulation was analyzed using the same cuts as in the analysis of

---
[19]Different cuts were applied, see Table 6.8

6.1. Analysis of simulated data

the experimental data [20]. The cuts are listed in Table 6.10. From *two body calculations* (section 5.1.2) information about the kinematics had been obtained, such as the maximum θ-angle of the ω-meson and the proton; these information were used to define appropriate cut settings.

In total 54200 events were reconstructed (after cuts) which led to an acceptance of:

$$Acc_{\omega \to \pi^0 \gamma} = 1.08\% \qquad (6.6)$$

Figure 6.27 shows the reconstructed invariant $e^+e^-\pi^0$-mass after cuts, and Figure 6.28 presents the corresponding two dimensional histogram with mass$_{e^-e^-}$ plotted against mass$_{e^-e^-\pi^0}$. In Figure 6.29 the invariant e^+e^--mass of the reconstructed events is shown. After division by the corresponding start distribution (Figure 5.18) the acceptance depending on the invariant mass$_{e^+e^-}$ was obtained, which is plotted in Figure 6.30. Again the shape of this histogram shows a *jagged* behavior at high invariant e^+e^--masses where the statistics becomes very low. The reason for this is the low statistics as well as the missing resolution effect in the start distribution[21]. Because of the results found in section 6.2.5 nothing was undertaken to correct this histogram.

Cut	Min	Max
BeamEnergy	1125.0	1410.0
θ-Proton	0.0	35.0
θ-omega	0.0	40.0
Momentum-X	-40.0	40.0
Momentum-Y	-40.0	40.0
Momentum-Z	-100.0	105.0
Missing Mass	910.0	975.0
Coplanarity	168.0	192.0
π^0 mass	110	160
e^- Cluster Size	5	15
e^+ Cluster Size	3	15

Table 6.10.: Applied cuts in the exclusive analysis of $\omega \to e^+e^-\pi^0$ (simulation).

6.1.9. SIM: Exclusive analysis of $\pi^0\eta$-production

The production of $\pi^0\eta$ pairs off the proton was also investigated. To determine the according cross section a simulation had to be performed, in order to obtain the acceptance. Thus 300.000 events were simulated for each of several intervals of incident

---
[20] Except for the cuts on the timing.
[21] Compare section 6.1.3.

6. Analysis

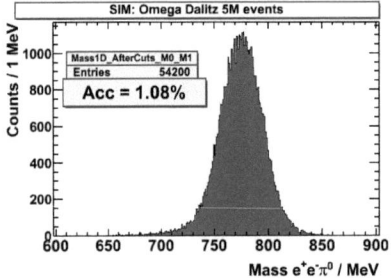

Figure 6.27.: Sim: invariant mass of reconstructed events in the analysis of $\omega \to e^+e^-\pi^0$.

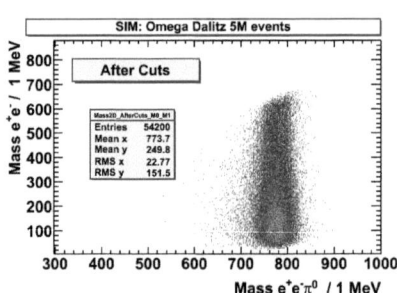

Figure 6.28.: Sim: 2D plot of mass$_{e^+e^-}$ versus the mass$_{e^+e^-\pi^0}$ in the analysis of the ω-Dalitz decay.

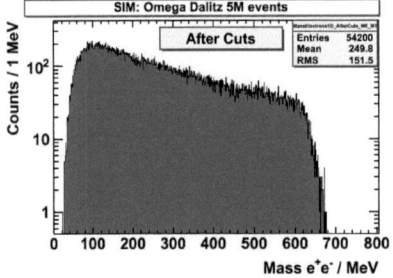

Figure 6.29.: Sim: distribution of the invariant e^+e^--mass of reconstructed ω-events.

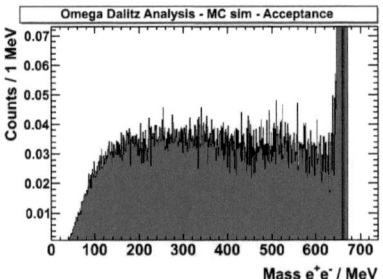

Figure 6.30.: Sim: acceptance of ω-Dalitz detection depending on m$_{e^+e^-}$.

energy; these intervals are listed in Table 6.12.

However, in the analysis of the simulated and the experimental data the cuts listed in Table 6.10 were applied. Before these cuts were applied, events with four photons and one proton were selected. A χ^2-test was used to identify the *best* pion. Table 6.11 lists the determined acceptances for the different intervals of incident energy. The Figures A.2 and A.3 (in the appendix) show the reconstructed invariant π^0 and η-masses for each interval.

6.1. Analysis of simulated data

Cut	Min	Max
Momentum-X	-40.0	40.0
Momentum-Y	-40.0	40.0
Momentum-Z	-100.0	105.0
Missing Mass	910.0	975.0
Coplanarity	168.0	192.0
Best Pion	110	160
Beam energy	min*	max*

Table 6.11.: Applied cuts in the analysis of $\pi^0\eta$. (*)The cut on the beam energy corresponds to each of the intervals (see Table 6.12).

Beam energy [MeV]	Acceptance [%]
935 - 985	15.5
985 - 1035	15.4
1035 - 1085	14.2
1085 - 1135	14.2
1135 - 1185	13.9
1185 - 1235	13.5
1235 - 1285	12.6
1285 - 1335	11.9
1335 - 1385	11
1385 - 1410	11

Table 6.12.: Determined Acceptance of $\pi^0\eta$ for different intervals of incident energy.

6.1.10. SIM: Exclusive analysis of $\pi^0 \rightarrow e^+e^-\gamma$

Besides the η-Dalitz and the ω-Dalitz decay the Dalitz decay of the π^0-meson ($\pi^0 \rightarrow \gamma\gamma* \rightarrow e^+e^-\gamma$) was also investigated. As the pion is very light in mass ($\approx$ 135 MeV), the coupling of a vector meson onto the $\gamma*$ is expected to be extremely small. Thus, the shape of the distribution of the invariant e^+e^--mass should correspond to the QED and the form factor is expected to be equal to one.

As in the investigation of the distribution of $m_{e^+e^-}$ in the experimental data an acceptance correction had to be applied, a simulation had to be performed. Five million events were generated using the PLUTO event generator and were simulated using *A2sim*. Thereafter the $AR_{HB}2v3$ analysis program was used to analyze the simulated data.

At first this investigation was done using a cluster threshold of 20 MeV. Unfortunately this led to difficulties concerning the invariant masses of e^+e^- in the regime of the kinematic limit, which could not be solved. The red ellipse in Figure 6.32 marks the problematic regime in the 2D-plot of $m_{e^+e^-}$ versus $m_{e^+e^-\gamma}$ (after cuts). The corresponding plot before cuts is shown in Figure 6.31.

Hence, a larger cluster threshold of 50 MeV was used to overcome these difficulties. Figure 6.33 shows the same 2D-plot after cuts using the higher threshold. Still, some light *spread* of the signal remains along the kinematic limit (which corresponds to the diagonal in the 2D-plot). The disadvantage of the higher cluster threshold is of course, that the statistics is reduced. It was found, that the detection efficiency (after cuts) dropped at least by a factor of ≈ 4. Figure 6.34 shows the invariant mass of $e^+e^-\gamma$ after

6. Analysis

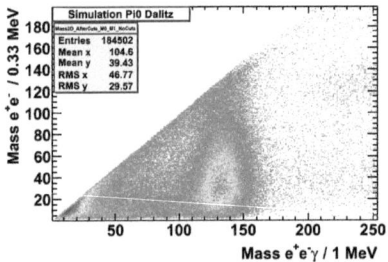

Figure 6.31.: Simulation of π^0-Dalitz using a cluster threshold of 20 MeV before cuts.

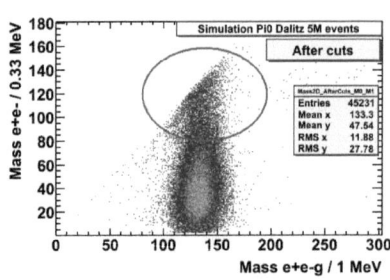

Figure 6.32.: Simulation of π^0-Dalitz using a cluster threshold of 20 MeV after cuts; the red ellipse marks the problematic regime.

the application of all cuts, which are listed in Table 6.13. The overall acceptance is 0.2 %, which is very small.

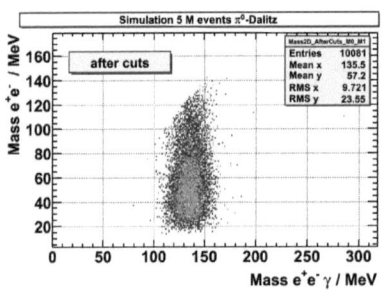

Figure 6.33.: Simulation of π^0-Dalitz; invariant mass$_{e^+e^-\gamma}$ of reconstructed reconstructed events (after cuts and threshold of 50 MeV).

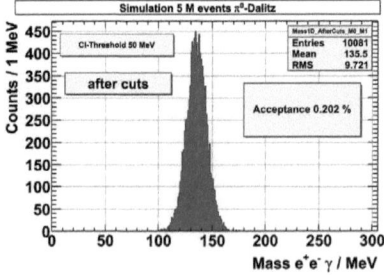

Figure 6.34.: Simulation of π^0-Dalitz using a cluster threshold of 50 MeV after cuts.

In order to obtain a specific acceptance of the invariant e^+e^--mass, the corresponding mass of the reconstructed events after cuts (Figure 6.35) was divided by $m_{e^+e^-}$ of the starting distribution (Figure 5.20). The result of this is shown in Figure 6.36. As one can see, the acceptance increases for higher invariant masses, which is not realistic. The reason for this rise is the remaining spread of the π^0-signal alongside the diagonal of the 2D-plot, which has been mentioned before. Thus more statistics is generated at higher invariant masses, where the corresponding statistics in the start distribution drops strongly. Hence, the acceptance had to be corrected. The red line in Figure 6.36 illustrates the distribution of the corrected acceptance[22].

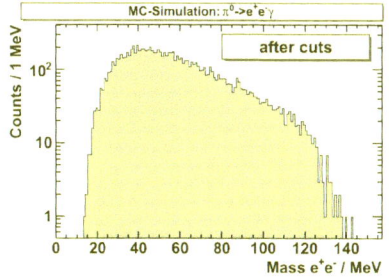

Figure 6.35.: Simulation of π^0-Dalitz; invariant mass$_{e^+e^-}$ of reconstructed events (after cuts).

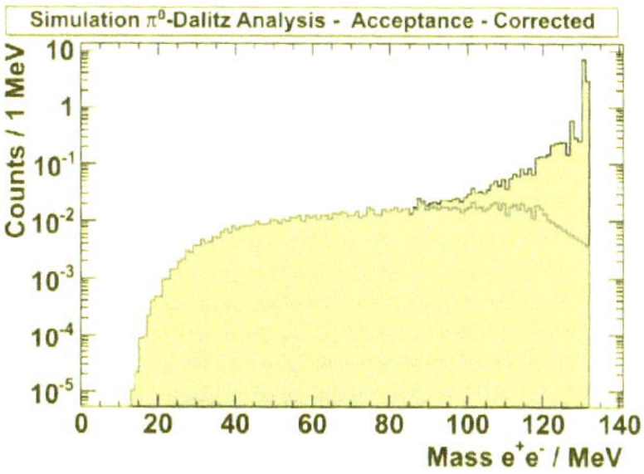

Figure 6.36.: Acceptance of the π^0-Dalitz decay; the red line shows the corrected shape of the acceptance (see text).

Cut	Min	Max
BeamEnergy	610.0	1410.0
θ-Proton	0.0	50.0
θ-omega	0.0	40.0
Momentum-X	-40.0	40.0
Momentum-Y	-40.0	40.0
Momentum-Z	-100.0	105.0
Missing Mass	910.0	975.0
Energy Balance	-40	40
Coplanarity	168.0	192.0

Table 6.13.: Applied cuts in the exclusive analysis of $\pi^0 \to e^+e^-\gamma$ (simulation).

6.2. Analysis of experimental data

In the following sections the analysis of the experimental data will be discussed. In principle the steps in the analysis are equal to the steps in the investigation of simulated data.

---
[22]This correction might only be accurate up to 125 MeV, which does not bother since the highest data point in the analyses of the experimental data is below 125 MeV.

6. Analysis

However, as far as the *experimental* data are concerned additional cuts on the prompt peaks in the timing spectra have to be applied. The Figures 6.37 to 6.38 show the timing spectra of the detected protons in the analysis of the η-Dalitz decay (for a fraction of the whole data; before and after cuts). As a particle can either be detected in TAPS or in the CB, but the timing is always filled into the same histogram, the timing-spectra contain both timings (CB-Tagger and TAPS-Tagger). In order to separate the TAPS timing signals from the CB-Timing signals, the former were shifted by a constant[23]. The Figures A.4 to A.5 (in the Appendix) show the timing spectra of the remaining particles in the analysis of the η-Dalitz decay.

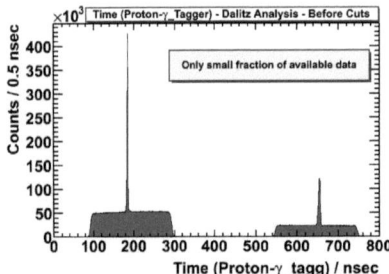

Figure 6.37.: Plot of the timing information of detected protons in the η-Dalitz analysis (before cuts).

Figure 6.38.: Plot of the timing information of detected proton in the η-Dalitz analysis (after cuts, except for the cut on the proton timing).

The width of time-windows on the prompt peaks in CB were $\approx 7ns$; in TAPS they were smaller due to the better timing resolution ($\approx 4nsec$). Further on cuts on the prompt peak in the timing spectrum of the TAGGER were applied (width $\approx 9nsec$). For each accomplished analysis the used cuts on the prompt peaks in the timing spectra are listed in extra Tables.

Moreover stricter cuts on variables such as the *time of flight*[24] (TOF) were applied during the analysis of the ntuples contained in the preselected data-files[25]. The TOF spectrum and the applied cut are shown in the Figures 6.39 and 6.40.

In all analyses it is of importance to investigate the background and to identify possible background channels, especially in analysis of the Dalitz decays. In section 6.4 an

---

[23] The shifting was accomplished in the moment the histograms were filled.
[24] Only possible for protons detected in TAPS.
[25] In this step of the analysis the program *HBAnalysis* was used for the analysis of the ntuples.

6.2. Analysis of experimental data

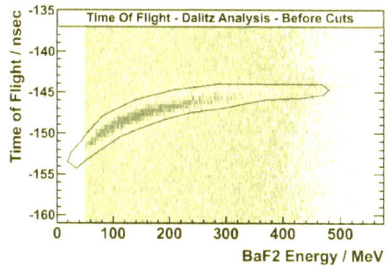

Figure 6.39.: Plot of the time of flight information of detected protons (TAPS) in the η-Dalitz analysis (before cuts).

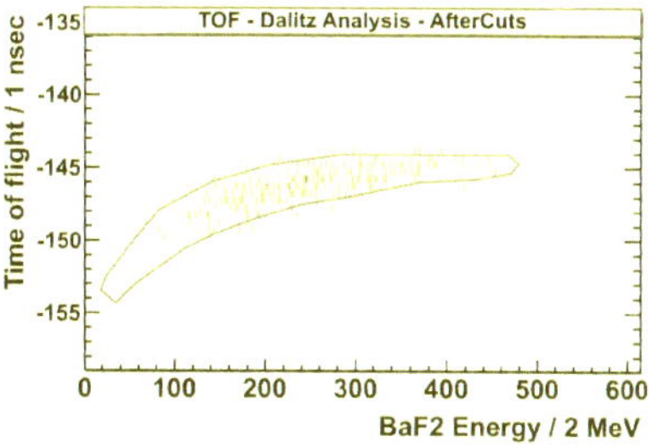

Figure 6.40.: Plot of the time of flight information of detected protons (TAPS) in the η-Dalitz analysis (after cuts). The used TOF-banana cut is illustrated by the red curve.

investigation of the background in the η-Dalitz analysis is presented

6.2.1. DATA: Exclusive analysis of $\eta \to \gamma\gamma$

In the exclusive analysis of $\eta \to \gamma\gamma$ the cuts listed in Table 6.1 were applied. Additional cuts on the prompt peaks in the timing spectra were used; these are listed in Table 6.14. All neutral multiplicities and all proton multiplicities were investigated. However, it was checked that non of the events entered more than once in the final results. Figure 6.41 shows the invariant $\gamma\gamma$-mass for the reconstructed η-events. In total 249500 events were reconstructed. The statistical error is: 499 counts.

Cut	min [ns]	max [ns]
Tagger Hit	-146.0	-137.0
Photons in CB	180.0	187.0
Protons in CB	180.0	187.0
Photons in TAPS	655.0	659.0
Protons in TAPS	648.0	658.0

Table 6.14.: Applied *time* cuts in the exclusive analysis of $\eta \to \gamma\gamma$.

With both, the acceptance determined in section 6.1.1 and the known branching ratio of $39.31 \pm 0.2\%$[26] the number of produced η-mesons in the data-set 2007-07-lH_2 was

---
[26] Particle Data Group 2008.

6. Analysis

determined as[27]:

$$N_{\eta_{\text{produced}}} = (5.23 \pm 0.26) \cdot 10^6 \qquad (6.7)$$

The error was calculated in the following way (C = counts, B = branching ratio and A is the acceptance[28]):

$$\triangle N = \sqrt{|\frac{\partial N}{\partial C}|^2 \cdot \triangle C^2 + |\frac{\partial N}{\partial B}|^2 \cdot \triangle B^2 + |\frac{\partial N}{\partial A}|^2 \cdot \triangle A^2}$$

$$\triangle N = \sqrt{|\frac{\triangle C}{B \cdot A}|^2 + |\frac{C \cdot \triangle B}{B^2 \cdot A}|^2 + |\frac{C \cdot \triangle A}{B \cdot A^2}|^2} \qquad (6.8)$$

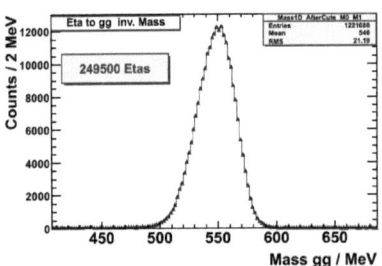

Figure 6.41.: DATA: reconstructed invariant $\gamma\gamma$-mass after cuts in the exclusive η-analysis.

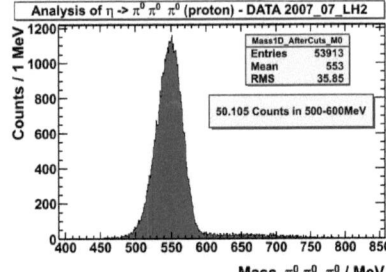

Figure 6.42.: DATA: reconstructed invariant $3\pi^0$-mass after cuts.

6.2.2. DATA: Exclusive analysis of $\eta \to \pi^0\pi^0\pi^0$

In the exclusive analysis of $\eta \to \pi^0\pi^0\pi^0$ the same analysis-function was used as for the corresponding investigation of the simulated data. Furthermore the same cuts were applied (Table 6.2). In addition cuts on the prompt peaks in the timing spectra of the detected particles were applied; these are listed in Table 6.14[29]. Only the data run 200-707-lH_2 was analyzed and higher multiplicities were not taken into account[30].

---

[27] According to: $N = Counts/(BR \cdot Acc)$.
[28] The absolute error of the acceptance, that was determined using a MC-simulation, was estimated as 5%. This is a common estimation of the used acceptances determiend with the *A2sim* program [26].
[29] In the analysis of $\eta \to 3\pi^0$ the same cuts on the prompt peaks in the timing spectra were applied as in the analysis of $\eta \to \gamma\gamma$
[30] Anyway, events containing more than eight hits in the CB/TAPS system occur rather seldom. Further on split-offs do not play a role, as the higher cluster-threshold of 50 MeV was used.

Figure 6.42 shows the result of this investigation. In the spectrum of the reconstructed invariant $3\pi^0$-mass 50.105 counts are contained within the mass region of 500 MeV to 600 MeV. The upper limit of the remaining background in this region was estimated as 250 counts[31]. Thus the final number of reconstructed η-mesons in this channel is 49855. Hence, with this number, the branching ratio[32] and the determined acceptance (section 6.1.2) this leads to the following number of produced η-mesons in the data:

$$N_{\eta_{produced}} = (5.57 \pm 0.28) \cdot 10^6 \tag{6.9}$$

The error was calculated corresponding to equation 6.8.

6.2.3. DATA: Exclusive analysis of $\eta \to \pi^0 \gamma \gamma$

The analysis of the decay $\eta \to \pi^0 \gamma \gamma$ is a difficult task. The two main reasons for this are the branching ratio of $BR = (4.4 \pm 1.5) \cdot 10^{-4}$, which is rather small compared to branching ratios of other η-decays, and the strong background from the $\pi^0 \pi^0$-reaction. As nearly all of these events end up in the same final state of four[33] photons, a very huge background has to be assumed. Furthermore the reaction of the $\pi^0 \eta$-production contributes also to the background[34]. Moreover it is not easy to remove this background, as all these reactions fulfill the cuts on the *energy balance*, *momentum balance* and the *missing mass*.

In this analysis no higher multiplicities were taken into account and only the beamtime 2007-07-*lH2* was investigated. All events containing exactly one proton and four photons were analyzed. In a first step the best pion was identified via a χ^2-test. Thereafter, in order to suppress $\pi^0 \pi^0$-background, the request was made, that the invariant mass of the remaining two photons was not within the range of the π^0-mass. The Figures 6.43 and 6.44 show the final results after cuts. The shoulder that emerges on the right hand side of the Gaussian distribution is located at 548 MeV on the invariant mass axis and contains several entries, which can be considered as candidates of the decay $\eta \to \pi^0 \gamma \gamma$. The background was fitted by a double Gaussian; one Gaussian for the background and the other Gaussian for the signal. The former is shown as dashed curve in the Figures 6.34 and 6.44 (black) and has fixed values for the width = 34.49 MeV, height = 101.8 Counts/8MeV and the mean = 518.9 MeV. Concerning the signal-Gaussian: the values of the width and the mean of the Gaussian were confined to: 13 MeV < width < 15

---

[31]This upper limit of the background was determined as follows: on the right hand side of the η-peak the average hight of the background is $\approx$ 25 counts. Thus the background in 500 MeV to 600 MeV can be easily estimated with: 25 Counts ·100 bins of 1 MeV = 250 background-counts.

[32]$BR_{\eta \to 3\pi^0} = (32.51 \pm 0.28)\%$, Particle Data Group, 2006.

[33]Branching ratio of $\pi^0 \to \gamma\gamma$ is 98.78% (PDG).

[34]Only in the case of $\eta \to \gamma\gamma$ (BR = 39.39%) (PDG).

6. Analysis

MeV and 13 MeV 546 < mean < 549 MeV). Only the hight of the Gaussian was varied (by the fit-function). The number of events was calculated as[35]:

$$N_{\eta_{\pi^0\gamma\gamma}} = \frac{7.33 \cdot 14.0 \cdot 2.35}{8} = 30.14 \pm 19.11 \qquad (6.10)$$

with the error calculated in the following way:

$$\Delta C = \frac{2.35}{binning} \cdot \sqrt{(Height \cdot \Delta Width)^2 + (Width \cdot \Delta Height)^2} \qquad (6.11)$$

$$= \frac{2.35}{8} \cdot \sqrt{(7.33 \cdot 0.5)^2 + (14.0 \cdot 4.64)^2} = 19.11$$

Unfortunately the errors are huge; still, within the errors this number corresponds to an simple estimation of expected counts. The number of produced η-mesons in this data set is $\approx$ 5.5 million (compare to section 6.2.2). With the branching ratio and an assumed acceptance of 1% (after cuts) the number of expected counts is 25.

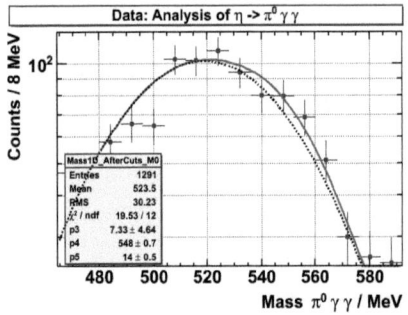 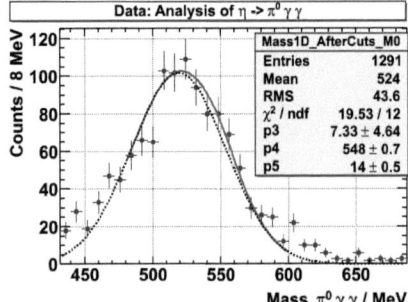

Figure 6.43.: DATA: reconstructed invariant $\pi^0\gamma\gamma$-mass after cuts in the exclusive analysis of $\eta \to \pi^0\gamma\gamma$. Red curve is the fit to the data (combination of two Gaussian functions).

Figure 6.44.: DATA: reconstructed invariant $\pi^0\gamma\gamma$-mass after cuts in the exclusive η-analysis. The dashed curve is the fit to the background (Gaussian).

6.2.4. DATA: Exclusive analysis of $\eta \to e^+e^-\gamma$

In the analysis of the η-Dalitz decay both beamtimes, 2007-06-lH2 and 2007-07-lH2 were analyzed. It was found that in the former run period the number of produced η-mesons

---
[35] The binning is 8. The error was calculated using equation 6.8, with the only difference that ΔC is given by the errors of the fit in Figure 6.44.

6.2. Analysis of experimental data

per data file was not stable and furthermore not comparable in total numbers to the latter beamtime. The reason for this remained unclear[36]. Still, as for the determination of the transition form factor of the η-Dalitz decay all that matters is statistics, both beamtimes were investigated with respect to higher multiplicities. Figure 6.45 shows the invariant mass of $e^+e^-\gamma$ before cuts.

Cut	min [ns]	max [ns]
Tagger Hit	-146.0	-137.0
Photons in CB	180.0	187.0
Protons in CB	180.0	187.0
Electrons in CB	180.0	187.0
Positrons in CB	180.0	187.0
Photons in TAPS	655.0	659.0
Protons in TAPS	648.0	658.0
Electrons in TAPS	655.0	659.0
Positrons in TAPS	648.0	658.0

Table 6.15.: Applied time cuts in the analysis of the data 2007-07–lH_2. Positrons refers to the 2nd charged hit, that is not a proton; electron to the first.

Cut	min [ns]	max [ns]
Photons in CB	172.0	181.0
Protons in CB	172.0	182.0
Electrons in CB	172.0	181.0
Positrons in CB	172.0	181.0
Photons in TAPS	655.0	659.0
Protons in TAPS	648.0	658.0
Electrons in TAPS	655.0	659.0
Positrons in TAPS	648.0	658.0

Table 6.16.: Applied time cuts in the analysis of the first fraction of data of the beamtime 2007-06–lH_2.

The analyses of the two data sets were accomplished separately. Later on both results were summed up and corrected by the acceptance correction, which is valid for both runs, as the run parameters of both beamtimes were equal and equal cuts were applied. The cuts are listed in Table 63. In addition cuts were applied on the prompt peaks in the timing spectra of the detected particles; these are listed in the Tables 6.15 and 6.16. Due to a shift of the prompt peak in the timing spectra of the Crystal Ball[37] after some hours of data taking, different time cuts had to be applied. The time shift was the result of a changed delay of the time-signal running between the tagger and the CB.

The spectra of the reconstructed invariant $e^+e^-\gamma$-mass as well as the 2D-plots of $m_{e^+e^-}$ versus $m_{e^+e^-\gamma}$ for both beamtimes are shown in the Figures 6.47 to 6.50. For the July beamtime the number of reconstructed η-mesons is 436 ± 31[38], which was determined by the fit in Figure 6.48. With the acceptance of 1.3% (section 6.1.3) and the known

---
[36] The author started with this analysis project after the data allocations in June/July 2007 were already accomplished by collaboration partners. It was found that the scalers did not work in the June beamtime. Thus, this run period can not be used in order to determine cross sections. Further on, as far as the measurement of branching ratios is concerned, the June beamtime can not be used, because the number of produced η-mesons was not stable.
[37] This is always the time-difference between the tagger and the Crystal Ball.
[38] The error was calculated according to equation 6.11.

6. Analysis

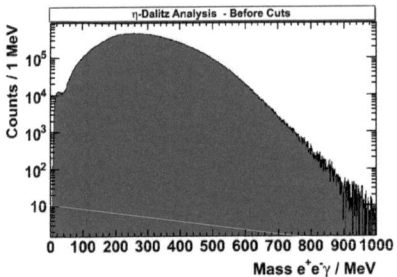

Figure 6.45.: DATA: reconstructed invariant $e^+e^-\gamma$-mass before cuts in the exclusive analysis of the η-Dalitz decay.

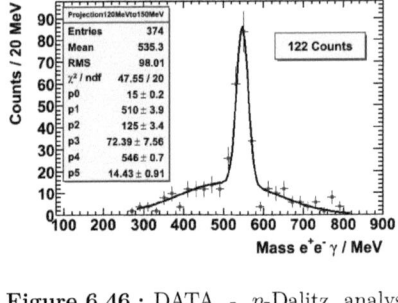

Figure 6.46.: DATA - η-Dalitz analysis: Projection of $m_{e^+e^-}$ onto $m_{e^+e^-\gamma}$ for the interval of 120 MeV to 150 MeV.

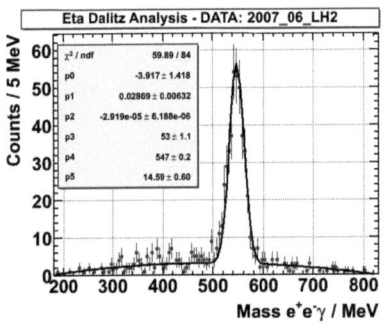

Figure 6.47.: DATA: reconstructed invariant $e^+e^-\gamma$-mass after cuts in the exclusive analysis of the η-Dalitz decay (Beamtime June 2007).

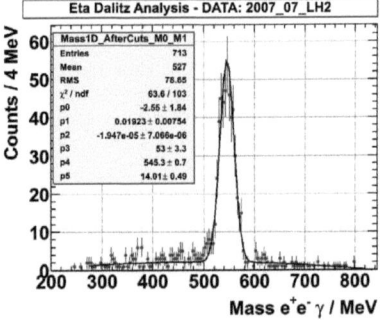

Figure 6.48.: DATA: reconstructed invariant $e^+e^-\gamma$-mass after cuts in the exclusive analysis of the η-Dalitz decay (Beamtime July 2007).

branching ratio[39] the number of in total produced η-mesons was determined as[40]:

$$N_{\eta_{\text{produced}}} = (5.58 \pm 0.88) \cdot 10^6 \qquad (6.12)$$

---

[39]$BR_{\eta \to e^+e^-\gamma} = (6.0 \pm 0.8) \cdot 10^{-3}$. (Particle Data Group 2006)
[40]The error was determined according to equation 6.8.

6.2. Analysis of experimental data

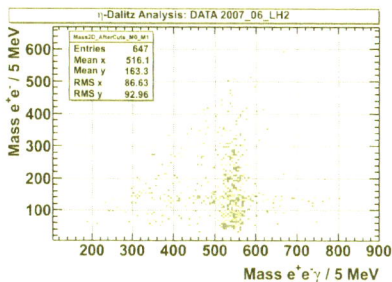

Figure 6.49.: DATA - η-Dalitz analysis: 2D-plot of the reconstructed invariant e^+e^--mass versus $m_{e^+e^-\gamma}$ after cuts (Beamtime June 2007).

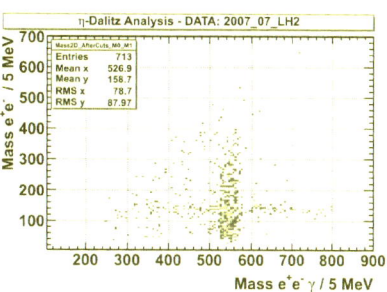

Figure 6.50.: DATA - η-Dalitz analysis: 2D-plot of the reconstructed invariant e^+e^--mass versus $m_{e^+e^-\gamma}$ after cuts (Beamtime July 2007).

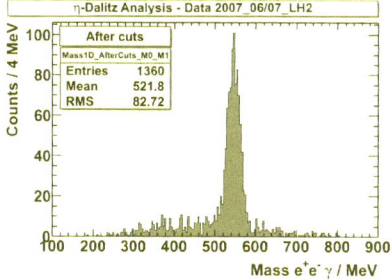

Figure 6.51.: DATA: reconstructed invariant $e^+e^-\gamma$-mass after cuts in the exclusive analysis of the η-Dalitz decay (both beamtimes added).

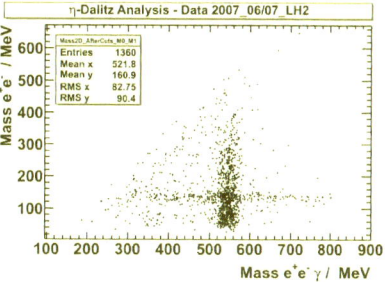

Figure 6.52.: DATA - η-Dalitz analysis: 2D-plot of the reconstructed invariant e^+e^--mass versus $m_{e^+e^-\gamma}$ after cuts (both beamtimes added).

Determination of the η transition form factor

In the next step the results of both beamtimes were summed up (Figure 6.51 and Figure 6.52). Thereafter projections were made onto the $m_{e^+e^-\gamma}$-axis for slices in $m_{e^+e^-}$ in steps of 30 MeV. These projections are shown in the Figures 6.53 and 6.54. Afterwards the η-signal region in each projection was integrated from 510 MeV to 590 MeV in order to determine the number of counts. As these numbers still contained background events, a background subtraction had to be accomplished.

This was realized by the application of the *side-band* subtraction method. Thus, in each of the projections the counts in left (430 MeV - 500 MeV) and in the right (600 MeV - 670

6. Analysis

MeV) side-band were integrated; the background counts of both bands were summed up and thereafter divided by a factor of two. Afterwards the result of this was subtracted from the counts in the signal region.
Table 6.17 lists the numbers of *counts without background*, *counts*, *background-counts*, etc.

This procedure of background-subtraction was used for all projections except for one. Only in the projection of the interval 120 MeV to 150 MeV enough counts were available, to fit the η-signal as well as the background.
As seen in the Figure 6.50 this interval is filled with a lot of misidentified π^0-events. As this band of misidentified π^0-mesons reaches into the η-signal region, the contribution of this background had to be determined thoroughly. The result of the fitting is shown in Figure 6.46. The number of η-Dalitz events in the mass region of 510 MeV to 590 MeV is 122.

After the subtraction of the background was accomplished, the invariant e^+e^--mass of the remaining η-signal counts was plotted (see Figure 6.55) and corrected by the corresponding acceptance (6.12 - section 6.1.3). The data points are listed in Table 6.17. The four data points for the highest invariant mass$_{e^+e^-}$ were merged into two points, each with a bin-width of 60 MeV. The QED-curve was scaled into the histogram (dashed curve). The result is shown in Figure 6.56; the black solid line is a fit to the data points using in addition a monopole form factor. Thereafter the measured data points were divided by the integral of the QED-curve in the corresponding interval of $m_{e^+e^-}$. The outcome of this procedure is shown in the two Figures 6.57 and 6.58 (see these Figures enlarged in chapter 7). The data points are listed in Table A.3. (appendix)

The red line in the Figures 6.57 and 6.58 is the fit to the measured data point. The result for the Λ-fit-parameter is:

$$\Lambda = 738.1 \pm 73.5 \text{ MeV}$$
$$\Lambda \approx 740 \pm 74 \text{ MeV} \tag{6.13}$$

Thus, for the slope parameter of the transition form factor of the η-meson follows:

$$b = \frac{dF}{dq^2}\Big|_{q^2=0} = \Lambda^{-2}$$
$$b = 1.836^{+0.428}_{-0.317} \frac{1}{\text{GeV}^2}$$
$$b \approx 1.84^{+0.43}_{-0.32} \frac{1}{\text{GeV}^2} \tag{6.14}$$

The result agrees with the prediction of Landsberg [31] as well as with the calculation provided by C. Terschluesen [46]. A further discussion about this result will be given in chapter 7.

6.2. Analysis of experimental data

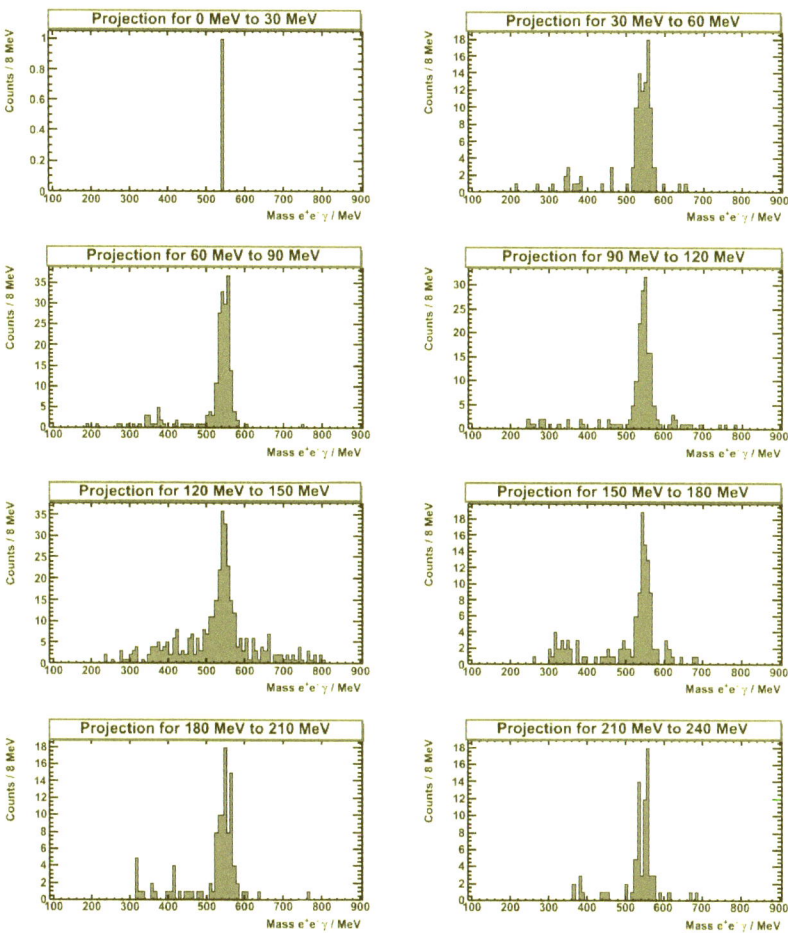

Figure 6.53.: η-Dalitz analysis: Projections of the 2D-plot of Figure 6.52 onto the $m_{e^+e^-\gamma}$-axis for slices in $m_{e^+e^-}$ with a width of 30 MeV.

Verification of applied cuts

One important issue was the verification of the cuts that were applied in the analysis. This was accomplished by plotting the corresponding cut-variable (e.g *missing mass*)

6. Analysis

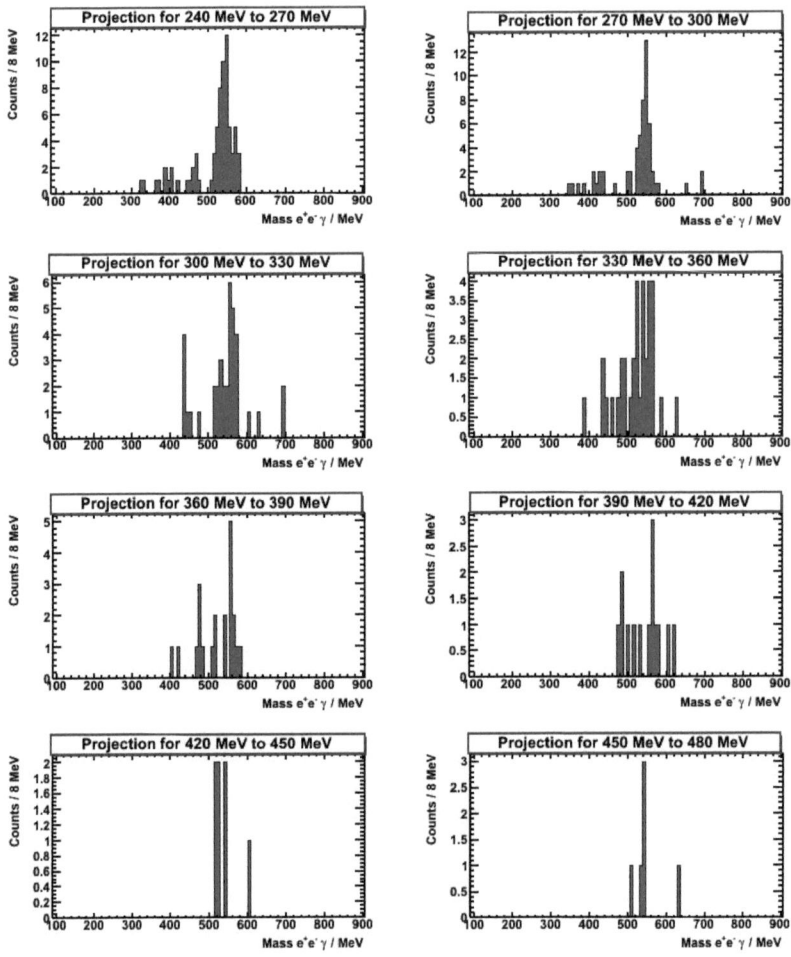

Figure 6.54.: η-Dalitz analysis: Projections of the 2D-plot of Figure 6.52 onto the $m_{e^+e^-\gamma}$-axis for slices in $m_{e^+e^-}$ with a width of 30 MeV.

after application of all other cuts listed in Table 6.3[41]. This procedure turned out to

---

[41]The only cut, that is NOT applied during this procedure is the cut on the *special* variable on which the focus of investigation was set (e.g *missing mass*).

6.2. Analysis of experimental data

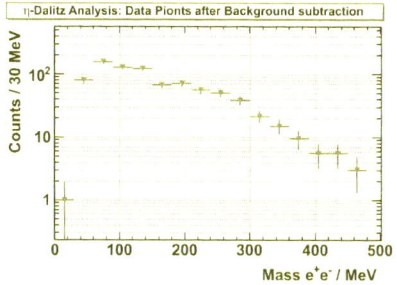

Figure 6.55.: DATA: reconstructed invariant e^+e^--mass after cuts and background subtraction without acceptance correction (both beamtimes 2007). See Table 6.17.

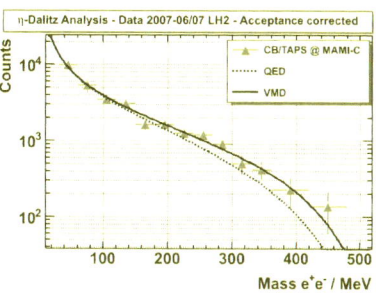

Figure 6.56.: DATA - η-Dalitz analysis: reconstructed invariant mass of e^+e^- after acceptance correction. The dashed black line is the scaled QED-curve. The black solid line is a fit to the data points within the VMD-model.

Mass [MeV]	Width [MeV]	Signal-Counts	Counts	Background	Acceptance
15.0	30	1.0	1.0	0.0	0.000
45.0	30	79.0	82.0	3.0	0.008012
75.0	30	157.0	162.00	5.0	0.029493
105.0	30	127.5	136.0	8.5	0.036352
135.0	30	122.0$^{(*)}$	(166)	(39.5)	0.039958
165.0	30	66.0	76.0	10.0	0.040195
195.0	30	71.0	75.0	4.0	0.043308
225.0	30	55.5	59.0	3.5	0.043804
255.0	30	38.0	40.0	2.0	0.042393
315.0	30	21.5	26.0	4.5	0.042764
345.0	30	15.0	20.0	5.0	0.035998
375.0	30	9.5	12.0	2.5	0.035048
405.0	30	5.5	8.0	2.5	0.030566
435.0	30	5.5	6.0	0.5	0.030923
465.0	30	3.0	4.0	1.0	0.028130

Table 6.17.: η-Dalitz Analysis: Data Points corresponding to Figure 6.55. (*) This data point is the result of the fit of the η-signal in the corresponding projection (see text).

be very helpful and indeed led to a verification of the cut-settings. E.g. if one looks

6. Analysis

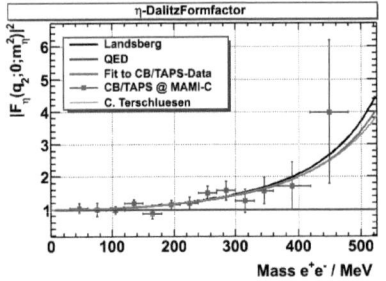
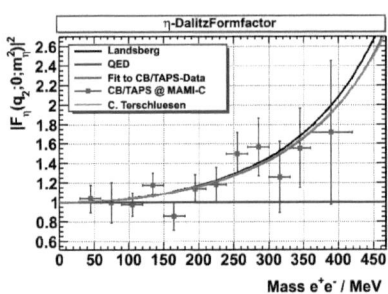

Figure 6.57.: DATA - η-Dalitz analysis: Transition form factor of the η-meson. The red curve is the fit to the data.

Figure 6.58.: DATA - η-Dalitz analysis: Transition form factor of the η-meson in comparison to theoretical predictions (zoomed view).

at the missing mass spectrum and applies all other cuts[42] (except for the missing mass cut itself), one can assume, that only those events with a more or less correct missing mass value survive in the η-Dalitz analysis. The Figures 6.59 and 6.60 illustrate this fact in case of the missing mass variable. In the latter Figure the cut on the missing mass (which is used in the η-Dalitz analysis) is illustrated by the blue lines.
The verification of all the cuts was done for both simulated data and experimental data. However, the results are presented here only for the verification of the experimental data.

In the same manner the variables *coplanarity* and *θ-proton* were investigated. The coplanarity has to be fulfilled for real η-Dalitz events; in other words: the value of the coplanarity value of these events has its peak at $\approx 180°$. Figure 6.61 shows the coplanarity after the application all cuts (except for the coplanarity cut itself). This result verifies the correctness of the chosen values of the cut range. The same holds for the θ-angle of the backscattered proton, which is shown in Figure 6.62 after the application of cuts. From *two body* simulations (section 5.1.1) the information was obtained, that the θ-angle of the proton has a maximum of $50°$ in the case of η-production off the nucleon with a maximum incident energy of 1.4 GeV. This was confirmed by the plot of the proton-θ-angle of events that survived the cuts[43].

The momentum balance in X direction of all events surviving the cuts[44] is shown in Figure 6.63. The corresponding Figure for the Y-direction looks alike and can be found in the appendix (Figure A.4) as well as a 2D-plot of the momentum balance in X direction

---

[42]In the authors work the missing mass was always the mass of a missing proton (no matter if it were detected or not).
[43]The cut on the θ-angle itself was not applied.
[44]Without cuts on the momentum balance variables.

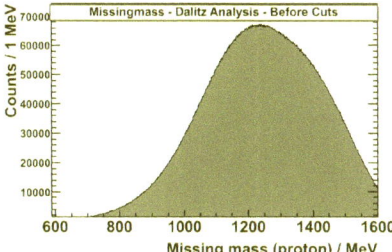

Figure 6.59.: Verification of the missing mass cut applied in the η-Dalitz analysis. Plot of the missing mass before cuts.

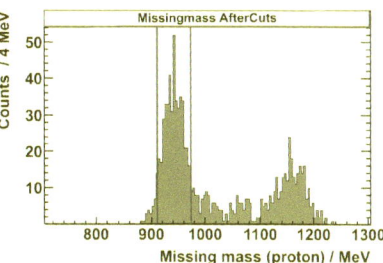

Figure 6.60.: Verification of the missing mass cut applied in the η-Dalitz analysis. Plot of the missing mass after all other cuts.

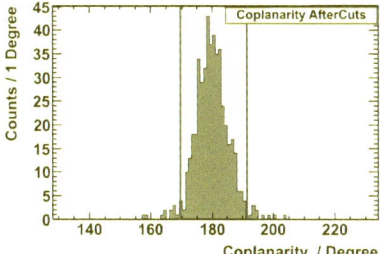

Figure 6.61.: Verification of the coplanarity cut applied in the η-Dalitz analysis. Plot of the coplanarity after all other cuts.

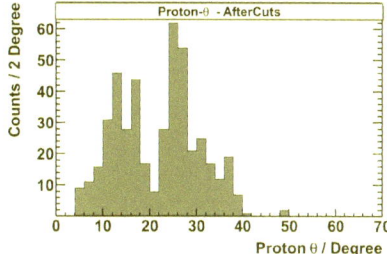

Figure 6.62.: Verification of the cut applied on the θ-proton in the η-Dalitz analysis. Plot of the θ-angle of the proton after all other cuts.

versus the energy balance (Figure A.17). In Figure 6.64 the momentum in Z direction is plotted after cuts.

The cluster sizes of the e^+e^--hits are shown in the Figures 6.65 and 6.66 (after cuts, except for the cuts on the cluster sizes). In the Dalitz analysis a cluster size ≥ 5 is required for the first hit, and a size of ≥ 3 is required for the second hit. The reason for the difference of the two spectra is given by the cluster algorithm, which processes the hits in order of the registered cluster energy; thus the first hit corresponds always to the particle with the highest energy deposit in an event. A higher energy deposit corresponds to a larger cluster size.

6. Analysis

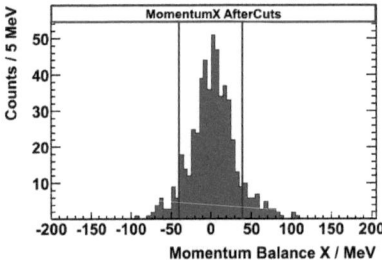

Figure 6.63.: Verification of the cut applied on *momentum balance X* in the η-Dalitz analysis. 1D-Plot of the corresponding variable after all other cuts.

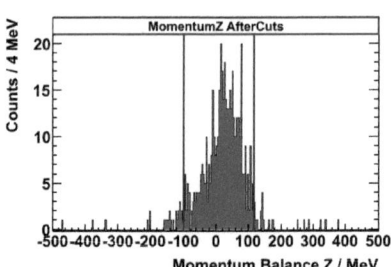

Figure 6.64.: Verification of the cut applied on *momentum balance Z* in the η-Dalitz analysis. Plot of the corresponding variable after all other cuts.

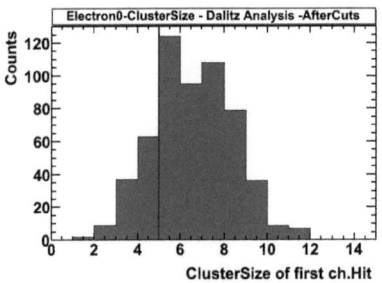

Figure 6.65.: Verification of the cut applied on *cluster size* in the η-Dalitz analysis. 1D-Plot of the corresponding variable after all other cuts.

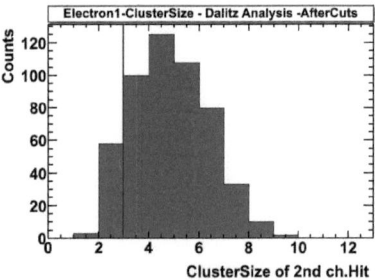

Figure 6.66.: Verification of the cut applied on *cluster size* in the η-Dalitz analysis. Plot of the corresponding variable after all other cuts.

The Figures 6.67 and 6.68 show the opening angle of the electron positron pair before and after the cuts (except for the cut on the opening angle itself). The lower and the upper limit of this cut are listed in Table 6.3. This cut helps to reduce background stemming from conversion processes[45] (section 6.4). See also Figure 6.102.

Cuts that helped to suppress the background very effectively are the ones applied on the

---

[45]The conversion of $\gamma \rightarrow e^+e^-$ can lead to a contribution to the background in the analysis of the η-Dalitz decay.

6.2. Analysis of experimental data

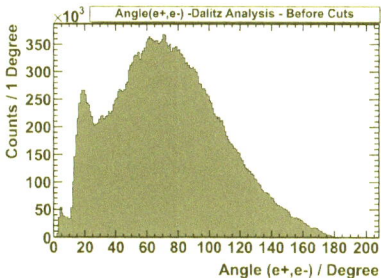

Figure 6.67.: Verification of the cut applied on the e^+e^--opening angle in the η-Dalitz analysis. 1D-Plot of the corresponding variable before cuts.

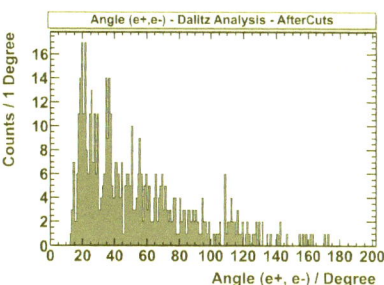

Figure 6.68.: Verification of the cut applied on the e^+e^--opening angle in the η-Dalitz analysis. Plot of the corresponding variable after all other cuts.

opening angles of $e^+\gamma$ and $e^-\gamma$. The corresponding histograms (before and after cuts) are shown in the Figures 6.69 and 6.70. A corresponding spectrum of the $e^+\gamma$-opening angle for background events (sideband 400 MeV to 500 MeV) is shown in Figure A.11 (appendix). A comparison to Figure 6.70 verifies the cut of 50°, as this cut removes much background and nearly no η-events.

All applied cuts were tested. The presented results demonstrate that the cut-settings are all appropriate.

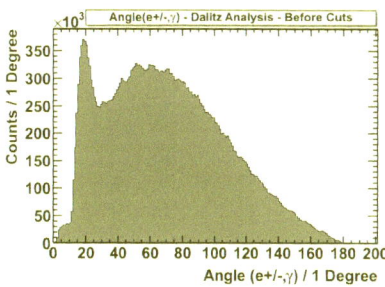

Figure 6.69.: Verification of the cut applied on the $e^{+/-}\gamma$-opening angle in the η-Dalitz analysis. 1D-Plot of the corresponding variable before all cuts cuts.

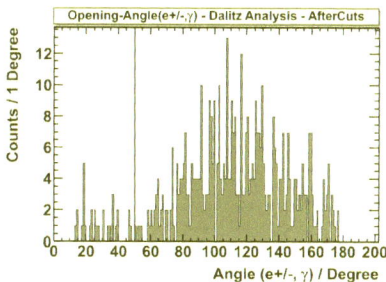

Figure 6.70.: Verification of the cut applied on the $e^{+/-}\gamma$-opening angle in the η-Dalitz analysis. Plot of the corresponding variable after all cuts cuts.

6.2.5. DATA: Exclusive analysis of $\eta \to \pi^+\pi^-\pi^0$

In the exclusive analysis of $\eta \to \pi^+\pi^-\pi^0$ all higher multiplicities with up to 6 $\pi^\pm$-hits, any number of detected protons and photons were investigated. In a first step only *kinematic*[46] cuts were applied, and thus not only an η-signal but also an ω-signal could be obtained. Figure 6.72 shows this result, whereas in Figure 6.71 the original spectrum of the invariant $\pi^+\pi^-\pi^0$-mass is plotted without any cuts applied. The *final* cuts that were applied are listed in Table 6.4. The additionally applied cuts on the prompt peaks in the timing spectra are listed in Table 6.18. The final result is shown in Figure 6.73.

In total 6956 ± 39 events[47] of $\eta \to \pi^+\pi^-\pi^0$ were reconstructed. With the branching ratio of $22.73 \pm 0.28\%$[48] and the acceptance determined in section 6.1.5 the number of in total produced η-mesons can be calculated[49]:

$$N_{\eta produced} = (5.46 \pm 0.28) \cdot 10^6 \qquad (6.15)$$

This result is in agreement with the number of η-mesons derived from the analysis of the other η-decay channels.

Cut	min [ns]	max [ns]
Tagger Hit	-146.0	-137.0
Photons in CB	180.0	187.0
Ch. Pions in CB	180.0	187.0
Protons in CB	180.0	187.0
Photons in TAPS	655.0	659.0
Ch. Pions in TAPS	655.0	659.0
Protons in TAPS	648.0	658.0

Table 6.18.: Applied *time* cuts in the exclusive analysis of $\eta \to \pi^+\pi^-\pi^0$.

6.2.6. DATA: Exclusive analysis of $\omega \to \pi^0\gamma$

In the exclusive analysis of $\omega \to \pi^0\gamma$ different cut settings were tested. In principle the signal to background ratio is more and more improved the stricter the cuts are. Thus, on the one hand strict cuts provide the means to obtain a *cleaner* signal, but on the other hand the statistics will be decreased by both, the strictness and number of the applied cuts.

---

[46]This refers to (loose) cuts on the missing mass, coplanarity and momentum balance.
[47]The error was determined using equation 6.11.
[48]Particle Data Group 2008.
[49]The error was determined according to equation 6.8.

6.2. Analysis of experimental data

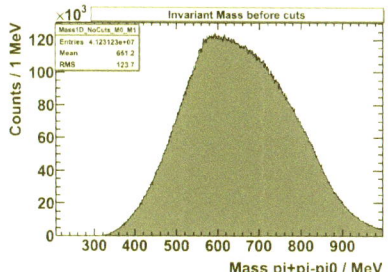

Figure 6.71.: Invariant mass of all detected $\pi^+\pi^-\pi^0$-events before cuts.

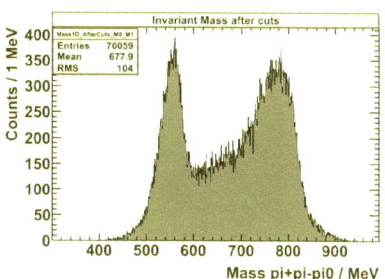

Figure 6.72.: Invariant mass of $\pi^+\pi^-\pi^0$-events after the application of cuts on the kinematics.

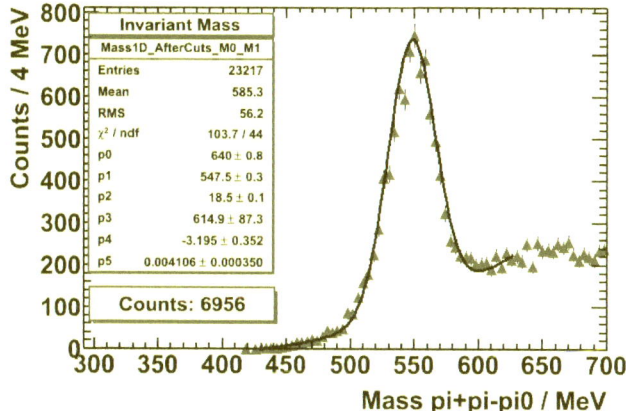

Figure 6.73.: Invariant mass of $\pi^+\pi^-\pi^0$-events after application of cuts listed in Table 6.4. The number of counts in the η-peak was determined by the fit to the data points.

In this investigation of the ω-meson only the beamtime 2007-07-LH_2 was analyzed taking higher multiplicities with any number of detected protons[50] and up to six photons

---

[50] This refers to all charged particles, that were reconstructed as protons. It is clear that in the investigated hadronic reaction only one proton can be involved; still: in the analysis class of $AR_{HB}2v3$ more than one proton per event can be available. The reason for this is the following: (e.g.) beside of the detection of the real backscattered proton a charged pion (or any other hit) can be additionally misidentified as proton.

6. Analysis

Cut	min [ns]	max [ns]
Tagger Hit	-146.0	-137.0
Photons in CB	180.0	187.0
Protons in CB	180.0	187.0
Photons in TAPS	655.0	659.0
Protons in TAPS	648.0	658.0

Table 6.19.: Applied *time* cuts in the exclusive analysis of $\omega \to \pi^0 \gamma$.

per event into account. Again it was ensured, that no event was counted more than once in the final histograms.

As other members[51] of the A2-Group of the University of Giessen investigated the $\pi^0\gamma$-decay of ω-mesons produced on heavy targets, similar cuts had to be applied in case of the analysis of the LH_2-data in order to allow for a comparison. Hence, the analysis was performed more than once testing different cuts on the π^0-mass. Only the results for the cuts $110 \leq m_{\pi^0} \leq 150$ and $128 \leq m_{\pi^0} \leq 143$ will be presented (for the Tables that list all applied cuts see section 6.1.5). They are shown in the Figures 6.74 to 6.78. In some of the presented plots only the mass region of interest is plotted; in these the background was fitted by a polynomial function of 3rd order and then subtracted from the data. In addition cuts on the prompt peak in the timing spectra were applied (Table 6.19).

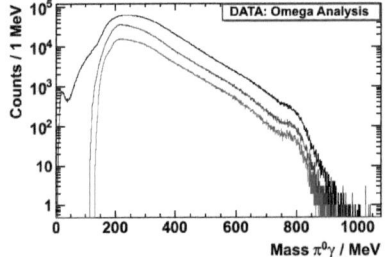

Figure 6.74.: Comparison of the invariant $\pi^0\gamma$ mass for different cuts (without time cuts).

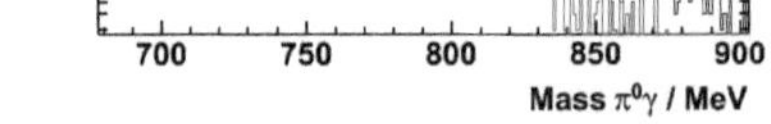

Figure 6.75.: Comparison of the invariant $\pi^0\gamma$ mass for different cuts (without time cuts).

Moreover an analysis without any time cuts was performed (Figures 6.74 and 6.75). This was done because of the following reason: the decay $\omega \to \pi^0 \gamma$ was also analyzed by

---

[51]M.Thiel and B. Lemmer[32].

members[52] of the CB/ELSA group of the University of Giessen for several target materials. As the Crystal Barrel detector (CB/ELSA) did not provide any timing information, a real comparison between the two experiments can only be realized, if the A2-data is analyzed without the cuts on the timing.

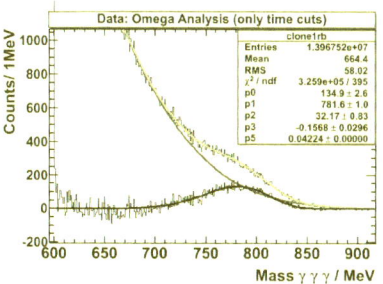 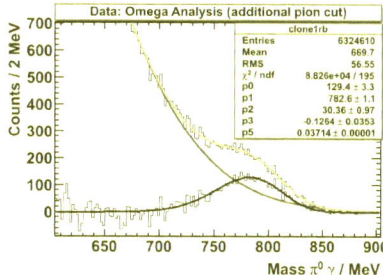

Figure 6.76.: Invariant $\gamma\gamma\gamma$-mass; only cuts on the timing were applied.

Figure 6.77.: Invariant $\pi^0\gamma$-mass; an additional cut on the π^0-mass was applied.

As has been mentioned before, stricter cut-settings and additional cuts (e.g on the kinematics) lead to a cleaner ω-signal. It was one further aim of the author to investigate several cut settings under the premise to achieve the best possible ω-signal to background ratio. The Figures 6.79, 6.80 and 6.81 show the results for three out of nine tested cut-settings. These cuts are listed in the Tables 6.5, 6.6, 6.7. The additional *time* cuts are listed in Table 6.19. In the analysis of simulated data, which was performed for each of the cut settings, the related detection efficiencies were determined. Thus, the number of determined counts in the ω-signal of the experimental data could be used in order to calculate the number of produced ω-mesons in the experiment. For this purpose the result was used, that was achieved after the application of the cut setting 'B' (Data: Figure 6.81, Sim: Figure 6.23). The number of ω-mesons in the peak has been determined as 690 with an error of 77[53]. With the known branching ratio of $BR = 8.9\%$ and the acceptance of 0.49% this leads to the following number of produced ω-mesons:

$$N_{\omega_{produced}} = (1.58 \pm 0.19) \cdot 10^6 \qquad (6.16)$$

The error in equation 6.16 was calculated corresponding to the equation 6.8.

---

[52] M. Nanova and K. Makonyi (both members of the 2. Physical Institute of the University of Giessen).
[53] Again the error was determined using the equation 6.11.

6. Analysis

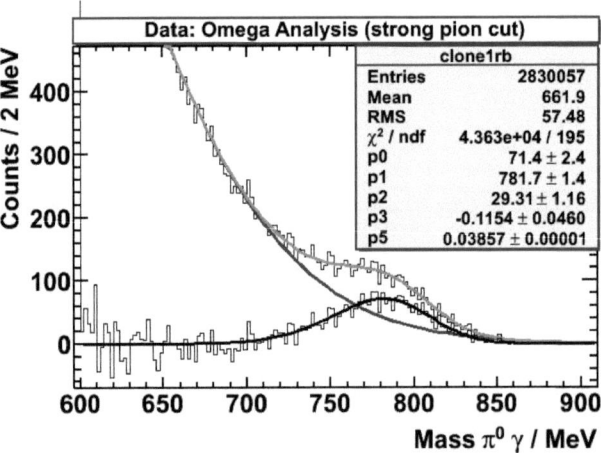

Figure 6.78.: Invariant $\pi^0\gamma$-mass after a strict cut on the π^0-mass (and cuts on the timing).

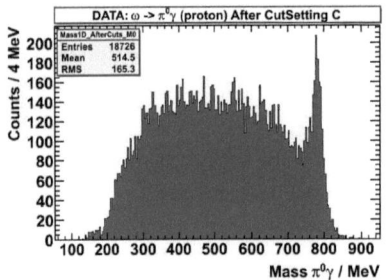

Figure 6.79.: Invariant $\pi^0\gamma$-mass after application of the cuts-list 'C'.

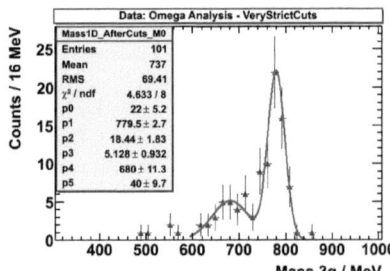

Figure 6.80.: Invariant $\pi^0\gamma$-mass after application of the cuts-list 'A'.

6.2.7. DATA: Exclusive analysis of $\omega \to \pi^+\pi^-\pi^0$

As the decay $\eta \to \pi^+\pi^-\pi^0$ had already been analyzed, an appropriate function of $AR_{HB}2v3$ for an analysis of the corresponding final state was available. Thus the analysis of $\omega \to \pi^+\pi^-\pi^0$ was accomplished rather fast. In the exclusive investigation of the experimental data[54] higher multiplicities were taken into account and the same cuts

---
[54] Only the beamtime 2007-07-LH_2 was analyzed.

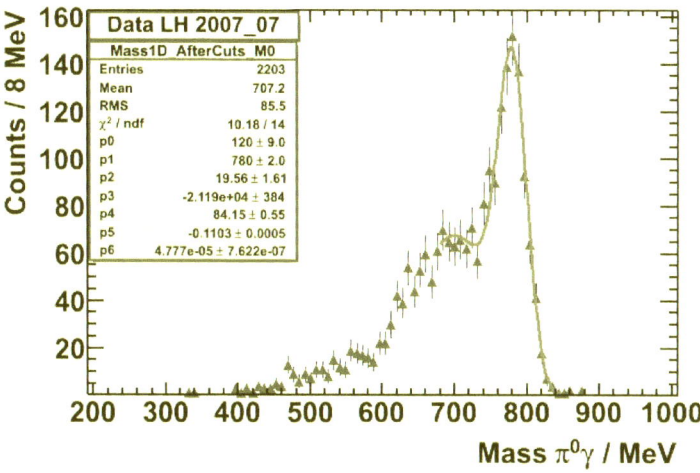

Figure 6.81.: Invariant $\pi^0\gamma$-mass after application of the cuts-list 'B'. The number of ω-mesons was determined by a fit to the data points.

were applied as for the simulated data (Table 6.8).

Figure 6.71 shows the invariant mass spectrum before cuts. In Figure 6.82 the final result is shown after the application of cuts. The remaining background underneath the ω-signal and the signal itself were fitted. The number of reconstructed ω-mesons (after cuts) is 8414 ± 138[55]. With this number, the determined acceptance (section 6.1.7) and the branching ratio[56] the number of ω-mesons produced in total was calculated as[57]:

$$N_{\omega_{produced}} = (1.576 \pm 0.083) \cdot 10^6 \qquad (6.17)$$

which is in a good agreement with the result obtained for the analysis of the $\pi^0\gamma$ decay channel.

6.2.8. DATA: Exclusive analysis of $\omega \to \pi^0 e^+ e^-$

Unfortunately the analysis of the ω-Dalitz decay was not promising due to the low statistics that were expected based on the information obtained through the investigations of

---
[55]The error was calculated according to equation 6.11.
[56]$BR = (89.1 \pm 0.7)\%$, Particle Data Group 2006.
[57]The error was calculated according to equation 6.8.

6. Analysis

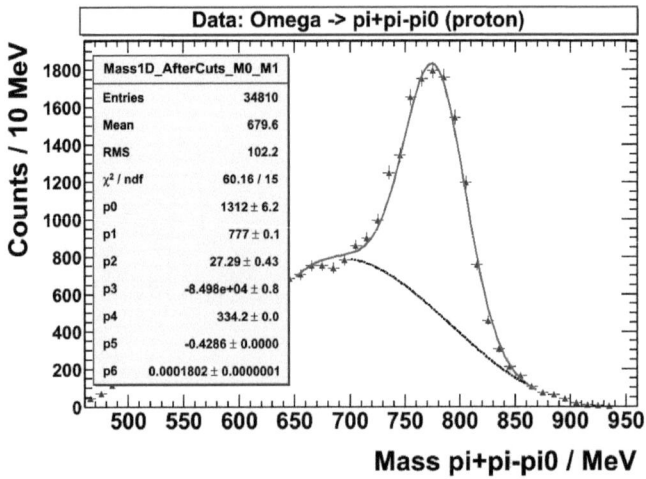

Figure 6.82.: Invariant $\pi^0\pi^+\pi^-$-mass after the application of cuts. The background and the ω-signal were fitted.

Cut	min [ns]	max [ns]
Tagger Hit	-146.0	-137.0
Photons in CB	180.0	187.0
Ch. Pions in CB	180.0	187.0
Protons in CB	180.0	187.0
Photons in TAPS	655.0	659.0
Ch. Pions in TAPS	655.0	659.0
Protons in TAPS	648.0	658.0

Table 6.20.: Applied *time* cuts in the exclusive analysis of $\omega \to \pi^+\pi^-\pi^0$.

other ω-decays. As the number of produced ω-mesons in the experiment[58] is roughly about 1.5 million, the expected number of detectable counts (after cuts) is approximately 12 counts. This number is the result of 1.5 million multiplied by the branching ratio[59] and the acceptance. The acceptance of the ω-Dalitz channel (after cuts) was determined as 1.08% (section 6.1.8), which is comparable to the acceptance in the case of the η-Dalitz decay.

As for the simulated data, the cuts listed in Table 6.9 were applied. Furthermore, cuts

---
[58]Only the beamtime 2007-07-LH_2
[59]$BR_{\omega \to e^+e^-\pi^0} = 7.7 \cdot 10^4$, Particle Data Group 2006.

on the timing spectra were applied and all higher multiplicities were analyzed. Figure 6.83 shows the resulting distribution of the invariant $e^+e^-\pi^0$-mass of all reconstructed events that survived the cuts. In Figure 6.84 the two dimensional plane plotting $m_{e^+e^-}$ versus $m_{e^+e^-\pi^0}$ is displayed.

It can not be claimed, that an ω-signal could be identified. Thus the result of the analysis of the ω-Dalitz decay can be simply stated as: In order to perform an analysis of the ω-Dalitz decay with convincing results, higher statistics are required. Hence, more ω-productions runs using a LH_2-target have to be performed in the future. As long as the maximum beam energy as well as the rates are not further increased, ten times more hours of beamtime are required to establish a basis for an appropriate investigation with the aim to reconstruct and identify an adequate number of counts in the $\pi^0 e^+ e^-$-final results.

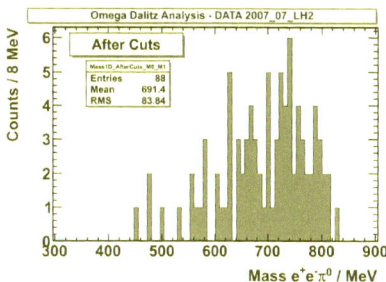

Figure 6.83.: Invariant mass of $e^+e^-\pi^0$ after all cuts.

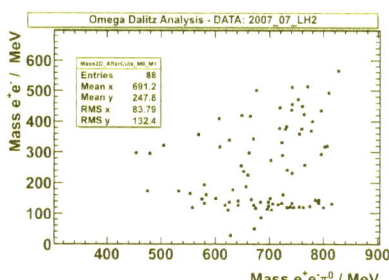

Figure 6.84.: 2D-plane of $m_{e^+e^-}$ plotted versus $m_{e^+e^-\pi^0}$ after cuts for the analysis of the ω-Dalitz decay.

6.2.9. DATA: Exclusive analysis of $\pi^0\eta$-production

In the first place, the analysis of $\pi^0\eta$-events was motivated by the idea to investigate the contribution of these events to the background in the analysis of $\omega \to \pi^0\gamma$. If one γ of the η[60] is not detected, which happens with a probability of $\approx 12\%$, the remaining particles end up in the same final state as the mentioned ω-decay. The same holds for $\pi^0\pi^0$-events. Hence, both reactions lead to an additional background in the ω-analysis, and because this analysis was an important aspect of the work of the A2 and CB/ELSA group, the decision was made to have a close look into these background channels.

---

[60]Only for events in which η decays into two photons.

6. Analysis

Cut	min [ns]	max [ns]
Tagger Hit	-146.0	-137.0
Photons in CB	180.0	187.0
Protons in CB	180.0	187.0
Photons in TAPS	655.0	659.0
Protons in TAPS	648.0	658.0

Table 6.21.: Applied *time* cuts in the exclusive analysis of $\pi^0\eta$-production.

In the analysis all events in the data[61] with 4 neutral hits and one proton were analyzed. In a first step the *best* possible combination of two photons giving a π^0 was identified by using a χ^2-test. In the second step the cuts listed in Table 6.10 were applied[62]; in addition the timing cuts listed in Table 6.21 were used. It was ensured, that for each detected η-count one π^0 count was present. Thereafter one of the photons of the reconstructed η was dropped at random, and the remaining three photons ($\pi^0\gamma$) were analyzed using the same function as for the investigation of $\omega \to \pi^0\gamma$. The result of this procedure led to the *background* shown in Figure 6.90.

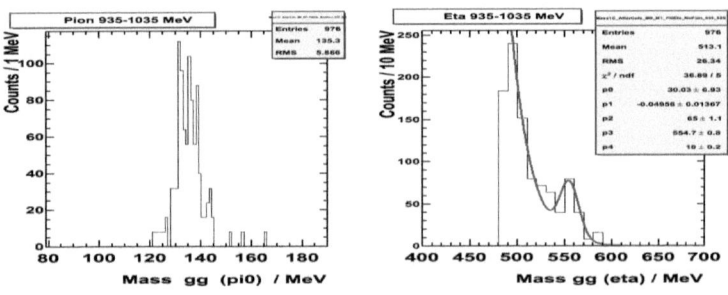

Figure 6.85.: Reconstructed π^0 and η-signals with a fit to the η-signal for the interval 935-1035 MeV of the incident energy.

Furthermore the cross section of $\pi^0\eta$-production was determined. In total ten intervals of the incident energy were investigated separately; these were 935 – 985 MeV, 985 – 1035 MeV, 1035 – 1085 MeV, 1085 – 1135 MeV, 1135 – 1185 MeV, 1185 – 1235 MeV, 1235 – 1285 MeV, 1285 – 1335 MeV, 1335 – 1385 MeV and 1385 – 1410 MeV. However, it was found that the statistics in the first two intervals was too low and thus these two intervals were merged into one. In the next step of the analysis, the η-signal was fitted in the

---
[61] In this respect, only the data set 2007-07LH_2 was analyzed.
[62] As the purpose of this investigation was to determine a possible background channel of $\omega \to \pi^0\gamma$, only events in the range of 1125-1400 MeV were analyzed.

6.2. Analysis of experimental data

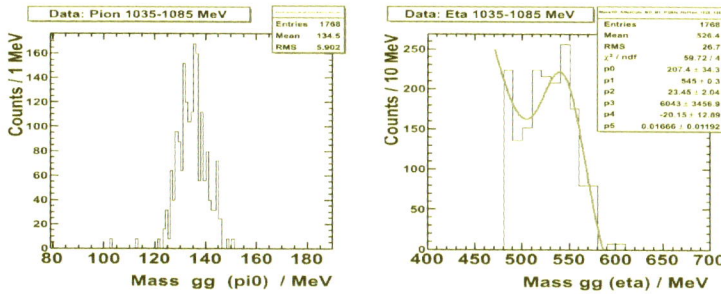

Figure 6.86.: Reconstructed π^0 and η-signals with a fit to the η-signal for the interval 1035-1085 MeV of the incident energy.

distribution of the invariant mass of those two photons, which did not make the *best* pion. It was ensured, that for each η-count a corresponding count in the π^0-mass region was detected. The π^0, η-signals are shown in the Figures 6.85 to 6.88 for the intervals with the lowest statistics and for an interval with higher statistics.

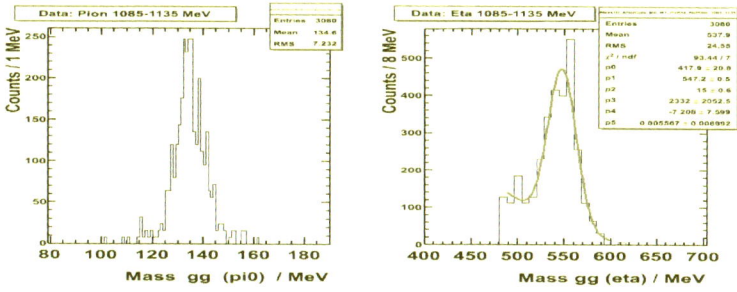

Figure 6.87.: Reconstructed π^0 and η-signals with a fit to the η-signal for the interval 1085-1135 MeV of the incident energy.

As the acceptance was already determined for each of the intervals in the analysis of the corresponding simulated data (Table 6.11), all that was required in addition for determining the cross section were the number of atoms in the target[63] and the *photon-flux*, which was determined by the author before. The determined values of the cross section, the acceptances as well as the *photon-flux* for each interval of the incident energy are listed in Table 6.22. The final result is shown in Figure 6.89. A comparison to a

---

[63]For the LH_2-target used in this experiment the value is: $N = 2 \cdot 10^{-7} \frac{1}{\mu b}$.

6. Analysis

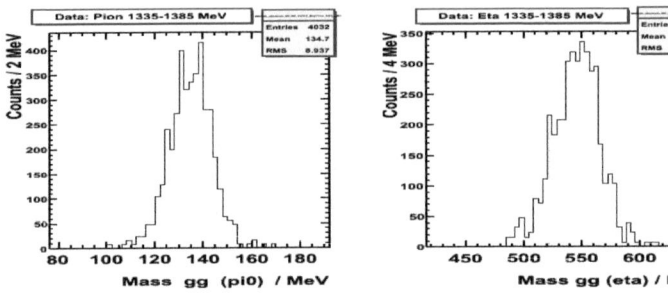

Figure 6.88.: Reconstructed π^0 and η-signals for the interval 1335-1385 MeV of the incident energy (without a plot of the fit).

published $\pi^0\eta$ cross section ([26], Appendix A.1) is given in chapter 7; the results agree within the errors.

Beam energy [MeV]	Width [MeV]	Counts	Flux	σ [μbarn]	$\Delta\sigma$ [μbarn]
985	100	152.75	1644315672576	0.0135	0.0013
1060	50	1143.48	727902322688	0.2504	0.0253
1110	50	1841.8	666395082752	0.4407	0.0445
1160	50	5144.0	660711342080	1.2682	0.1281
1210	50	7515.0	628080967680	2.0067	0.2027
1260	50	9872.0	611098034176	2.9029	0.2932
1310	50	7600.0	581577998336	2.4864	0.2511
1360	50	8288.0	493246611456	3.4587	0.3493
1410	25	3828.0	197287116800	3.9939	0.4034

Table 6.22.: Results of the investigation of the $\pi^0\eta$ production cross section. The acceptance for each interval is listed in Table 6.11.

6.2.10. DATA: Exclusive analysis of $\pi^0\pi^0$-production

In principle the analysis of $\pi^0\pi^0$-events was performed in the same manner as the analysis described in section 6.2.9, with the only difference, that the purpose was only to determine the contribution to the background in the analysis of $\omega \to \pi^0\gamma$ and $\eta \to e^+e^-\gamma$.

In order to investigate the $\pi^0\pi^0$-background in the $\pi^0\gamma$-analysis the cuts listed in Table 6.10 were applied and again a χ^2-test was used to determine the *best* pion. An additional π^0-cut was applied on the invariant mass of the other two photons. Thereafter, one out

6.2. Analysis of experimental data

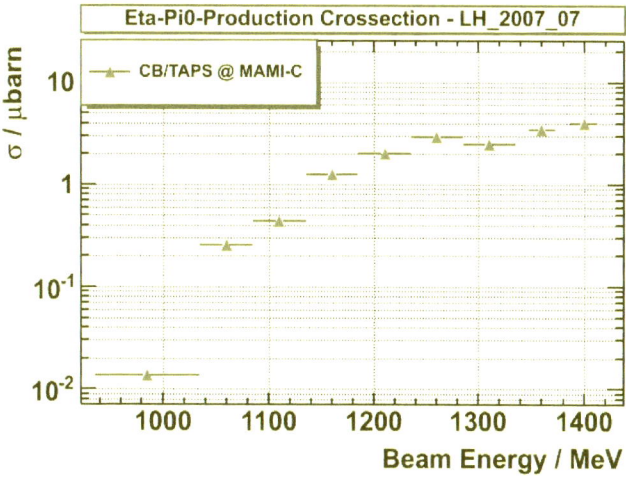

Figure 6.89.: Shown is the measured cross section for $\pi^0\eta$-production (result of this work). The data points are listed in Table 6.22.

of the two photons, that formed the second π^0, was dropped and the invariant $\pi^0\gamma$-mass was calculated. In Figure 6.90 the result of this procedure (and the $\pi^0\eta$-analysis accomplished in this context) is presented.

It was found that a strong source of background in the analysis of $e^+e^-\gamma$ are the $\pi^0\pi^0$-events. In the 2D-planes shown in the Figure 5.49 and 5.50 a background-band of pions that were misidentified ($\pi^0 \rightarrow \gamma\gamma \rightarrow e^+e^-$) can be seen (after cuts). It has been found in a simulation (section 6.4.1), that this band of misidentified pions mainly stems from $\pi^0\pi^0$-events. In order to estimate the probability for a misidentification of a pion that contributes the background in the η-Dalitz analysis after cuts, the number of produced $\pi^0\pi^0$-events in the data has to be determined in the first place. This was done using the cuts listed in Table A.4 (appendix) and Table 6.21; thereby the same incident energy range was used as in the η-Dalitz analysis. It can be seen in Figure A.19 (appendix) that 235000 events were reconstructed (after cuts). Using a Monte Carlo simulation the acceptance was determined as 4 % (Figure A.20). Thus the number of produced $\pi^0\pi^0$-events was calculated as:

$$N_{\pi^0\pi^0-produced} = \frac{235000}{0.04} \approx 5.88 \cdot 10^6 \quad \text{(beamtime July 2007)}$$

6. Analysis

This number is consistent with the fact that the cross section for η and for $\pi^0\pi^0$-production are of similar size[64] and that the number of produced η-mesons in the data was determined as 5.43 million (equation 6.26).
If all entries in the π^0-band in the Figure 5.50, that shows the 2D-plane of $m_{e^+e^-}$ versus $m_{e^+e^-\gamma}$ after cuts (beamtime July 2007), would stem from $\pi^0\pi^0$-events, the probability for a π^0-meson, that decays into two photons, to be misidentified and to contribute to the background in the η-Dalitz analysis after cuts could be calculated as follows:

$$P_{\pi^0 \to e^+e^-} = \frac{\text{Number of entries in the band}}{\text{Number of produced } \pi^0\pi^0} \quad \text{(after cuts)}$$

The number of entries in the band (Figure 5.50), was determined as $\approx$ 100 counts. This result was obtained in two steps; first a projection[65] of $m_{e^+e^-\gamma}$ onto the $m_{e^+e^-}$-axis was made. Thereafter the π^0-signal was fitted in order to determine the counts (Figure A.22 - appendix). Thus the probability for a pion to be misidentified ($\pi^0 \to \gamma\gamma \to e^+e^-$) and to contribute to the η-Dalitz analysis after cuts is given by:

$$P_{\pi^0 \to e^+e^-} = \frac{100}{5.88 \cdot 10^6} = 1.7 \cdot 10^{-5} \quad \text{(after cuts)}$$

This result is based on the assumption, that the $\pi^0\pi^0$-production is the only source for the misidentified pion background band in Figure 5.50, which is not absolutely correct, as the $\pi^0\eta$-production as well as the $\omega \to \pi^0\gamma$ decay can also contribute (section 6.4.1). Hence, the obtained result has to be seen as an upper limit for $P_{\pi^0 \to e^+e^-}$.

6.2.11. DATA: Exclusive analysis of $\pi^0 \to e^+e^-\gamma$

For the exclusive analysis of the π^0-Dalitz decay the same analysis function of $AR_{HB}2v3$ was used as in the case of the investigation of the η-Dalitz decay[66]; and thus, higher multiplicities were investigated again. Based on the information from the simulation of the π^0-Dalitz decay the investigation of the experimental data was performed using a cluster threshold of 50 MeV.

In this analysis the background correction was performed using the side band subtraction method. Hence, after the application of cuts, which are listed in the Tables 6.3 and 6.15[67], the remaining events in the 2D-plane shown in Figure 6.92 were projected onto the invariant e^+e^--mass-axis. This was done for three neighboring intervals corresponding to the *left side band* (50 MeV - 100 MeV), the *right side band* (180 MeV - 230 MeV), and the *signal region* (110 MeV - 160 MeV). Figure 6.93 shows the result of

---
[64]Range of incident photon energies: 750 MeV to 1210 MeV.
[65]In this projection the η-signal region was excluded.
[66]Because both decays end up in an equal final state ($e^+e^-\gamma$ (proton)).
[67]In the analysis of the π^0-Dalitz decay the same cuts on the prompt peaks in the timing spectra of the detected particles were used as in the analysis of the η-Dalitz decay.

6.2. Analysis of experimental data

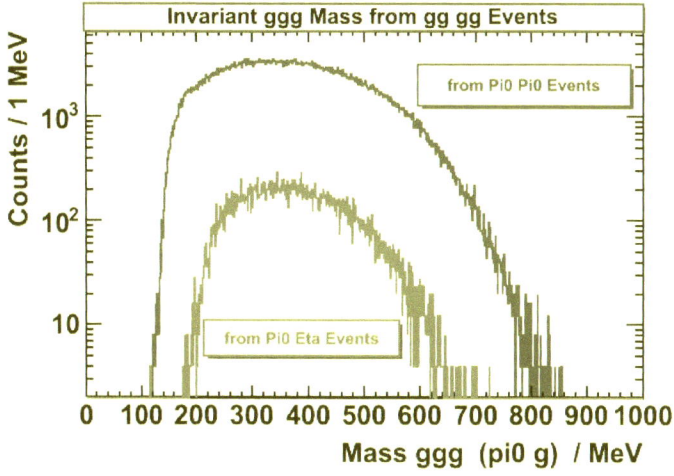

Figure 6.90.: Invariant $\pi^0\gamma$-mass stemming from $\pi^0\pi^0$ and $\pi^0\eta$-events.

this procedure. The black solid line is the distribution of $m_{e^+e^-}$ of the signal region, the green and the orange lines show the corresponding distributions of the two side bands. The side band distributions were summed up and normalized by a factor of 2.0 (red line - total background). Thereafter this sum was subtracted from the signal distribution; thus, the background subtraction was accomplished. The distribution of the remaining events within the signal region is shown by the blue solid line.

The number of reconstructed events was determined by a fit (red line in Figure 6.91). With a height of 145.7 counts/5 MeV, a width of 8.104 MeV and the factor 2.35, the number of counts was determined as:

$$C_{\pi^0-Dalitz} = 555 \pm 21$$

The error was calculated using the equation 6.11.

In order to perform an acceptance correction, the corrected distribution of $m_{e^+e^-}$ was divided by the acceptance (section 6.1.10). The result of this procedure is shown in Figure 6.94. As can be seen in Figure 6.93 the statistics is small for low invariant e^+e^--masses; thus the first two data points are uncertain. As can be seen in Figure 6.94 the distribution of the invariant e^+e^--mass stemming from the π^0-Dalitz decays does not differ from the QED, as expected. The data points are listed in Table 6.23. A further discussion of this result is given in chapter 7.

6. Analysis

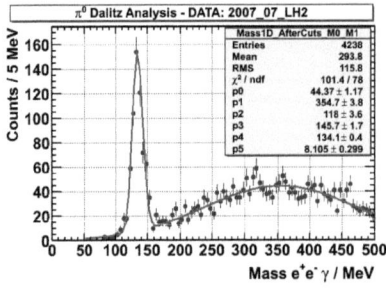

Figure 6.91.: π^0-Dalitz analysis: invariant $e^+e^-\gamma$-mass of the reconstructed events after cuts with a fit to the π^0-signal.

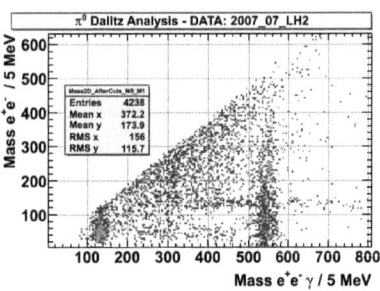

Figure 6.92.: Dalitz analysis: 2D-plot of $m_{e^+e^-}$ against $m_{e^+e^-\gamma}$ after cuts.

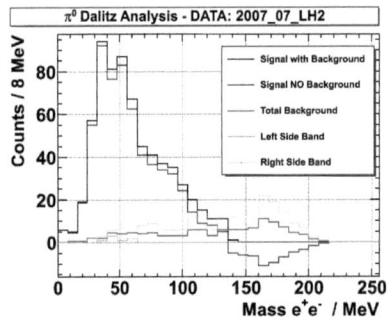

Figure 6.93.: π^0-Dalitz analysis: projection onto the $m_{e^+e^-}$-axis for the signal (black), the side bands (red, green), and the corrected signal (blue).

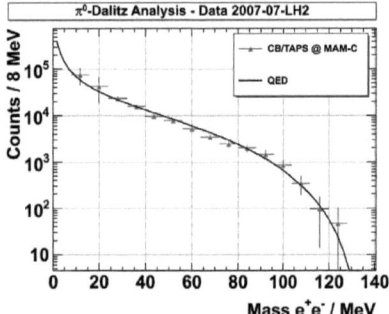

Figure 6.94.: π^0-Dalitz analysis: distribution of the invariant e^+e^--mass after background subtraction and after acceptance correction.

6.3. Determination of branching ratios

As in this work several decays of the η-meson and some decays of the ω-meson were investigated and the corresponding acceptances were determined, the different decay channels could be compared to each other. In principle the decay ratios can be tested according to:

$$\frac{\Gamma_{\eta \to \text{Decay-A}}}{\Gamma_{\eta \to \text{Decay-B}}} = \frac{\text{Counts}_A/\text{Acceptance}_A}{\text{Counts}_B/\text{Acceptance}_B} \qquad (6.18)$$

6.3. Determination of branching ratios

Mass [MeV]	Counts	After Acc. Correction	Error
12	4.5	79781.00	29963.47
20	18.5	40822.55	24.23
28.0	55	23591.15	1279.95
36.0	92	15568.13	1039.76
44.0	76.5	9054.41	792.95
52.0	83	7774.87	734.79
60.0	62.5	5158.70	598.53
68.0	41	2993.70	455.95
76.0	36.5	2435.06	411.21
84.0	34	2001.16	372.78
92.0	32	1476.06	320.16
100.0	24	854.79	243.64
108.0	14	345.95	154.99
116.0	9	95.61	81.48
124.0	8	46.66	56.92

Table 6.23.: π^0-Dalitz analysis: data points.

In this manner the intensity ratios of the channels $\eta \to \pi^+\pi^-\pi^0$, $\eta \to \pi^+\pi^-\pi^0$ and $\eta \to e^+e^-\gamma$ to the $\eta \to \gamma\gamma$ channel were calculated and compared to the ratio using the corresponding PDG values. In Table 6.24 the numbers of the reconstructed events of each analysis as well as the corresponding acceptances are listed. Using these numbers leads to the following ratios:

$$\frac{\Gamma_{\eta \to \pi^+\pi^-\pi^0}}{\Gamma_{\eta \to \gamma\gamma}} = 0.592 \pm 0.043 \tag{6.19}$$

$$\text{PDG: } \frac{\Gamma_{\eta \to \pi^+\pi^-\pi^0}}{\Gamma_{\eta \to \gamma\gamma}} = 0.576 \pm 0.011$$

$$\frac{\Gamma_{\eta \to \pi^0\pi^0\pi^0}}{\Gamma_{\eta \to \gamma\gamma}} = 0.879 \pm 0.062 \tag{6.20}$$

$$\text{PDG: } \frac{\Gamma_{\eta \to \pi^0\pi^0\pi^0}}{\Gamma_{\eta \to \gamma\gamma}} = 0.825 \pm 0.008$$

$$\frac{\Gamma_{\eta \to e^+e^-\gamma}}{\Gamma_{\eta \to \gamma\gamma}} = 0.0163 \pm 0.0016 \tag{6.21}$$

$$\text{PDG: } \frac{\Gamma_{\eta \to e^+e^-\gamma}}{\Gamma_{\eta \to \gamma\gamma}} = 0.0152 \pm 0.0020$$

$$\tag{6.22}$$

6. Analysis

For the PDG values the errors were calculated in the following way:

$$R = \frac{\Gamma_A}{\Gamma_B} \; ; \quad \Delta R = \sqrt{|\frac{\partial R}{\partial \Gamma_B}|^2 \cdot \Delta \Gamma_B^2 + |\frac{\partial R}{\partial \Gamma_A}|^2 \cdot \Delta \Gamma_A^2} \quad (6.23)$$

$$\Delta R_{PDG} = \sqrt{|\frac{\Gamma_A}{\Gamma_B^2}|^2 \cdot \Delta \Gamma_B^2 + |\frac{1}{\Gamma_B}|^2 \cdot \Delta \Gamma_A^2} \quad (6.24)$$

The error of each acceptances, which were determined by MC-simulations, was estimated with 5%[68]. Thus, with $R_{exp} = \frac{C_A \cdot Acc_B}{C_B \cdot Acc_A}$ the errors of the *experimental* values can be calculated as follows:

$$\Delta R_{exp} = \sqrt{|\frac{\partial R}{\partial C_B}|^2 \cdot \Delta C_B^2 + |\frac{\partial R}{\partial C_A}|^2 \cdot \Delta C_A^2 + |\frac{\partial R}{\partial Acc_B}|^2 \cdot \Delta Acc_B^2 + |\frac{\partial R}{\partial Acc_A}|^2 \cdot \Delta Acc_A^2}$$

$$= \sqrt{|\frac{C_A \cdot Acc_B}{C_B^2 \cdot Acc_A}|^2 \cdot \Delta C_B^2 + |\frac{Acc_B}{C_B \cdot Acc_A}|^2 \cdot \Delta C_A^2}$$

$$\cdot \sqrt{|\frac{C_A}{C_B \cdot Acc_A}|^2 \cdot \Delta Acc_B^2 + |\frac{C_A \cdot Acc_B}{C_B \cdot Acc_A^2}|^2 \cdot \Delta Acc_A^2} \quad (6.25)$$

Decay	Counts	Acceptance [%]	Branching ratio [%]
$\eta \to \gamma\gamma$	249500 ± 499	12.1	39.38 ± 0.26
$\eta \to \pi^0\pi^0\pi^0$	49855 ± 223	2.75	32.51 ± 0.28
$\eta \to \pi^+\pi^-\pi^0$	6956 ± 39	0.56	22.7 ± 0.4
$\eta \to e^+e^-\gamma$	436 ± 31	1.3	0.6 ± 0.08
$\eta \to \pi^0\gamma\gamma$	30 ± 19	–	0.04 ± 0.016
$\omega \to \pi^0\gamma$	690 ± 77	0.43	$8.9^{+0.27}_{-0.23}$
$\omega \to \pi^0\pi^+\pi^-$	8414 ± 138	0.599	89.1 ± 0.7

Table 6.24.: List of analyzed decays in the July beamtime, reconstructed events and the corresponding acceptances; the values of the branching ratios were taken from [20].

In the determination of a branching ratio, the number of produced mesons and the number of reconstructed mesons enters. Concerning the η-meson the former number was determined in the analysis of each investigated decay channel (using the PDG value of the corresponding branching ratio). The decays $\eta \to \gamma\gamma$ and $\eta \to \pi^0\pi^0\pi^0$ are the strongest η-decays (with the highest branching ratios); thus the statistics in the analysis of these decays was comparatively higher. In order to determine a reliable number of produced mesons, the arithmetic mean of the corresponding numbers obtained by the

---
[68]Which is a common estimation used by other collaboration partners as well [26].

6.3. Determination of branching ratios

analysis of $\eta \to \gamma\gamma$ and $\eta \to \pi^0\pi^0\pi^0$ was calculated:

$$\begin{aligned} N_{\eta_{\text{produced}}} &= \frac{N_{\gamma\gamma\text{-Analysis}} + N_{3\pi^0\text{-Analysis}}}{2} \\ &= \frac{(5.23 + 5.57) \cdot 10^6}{2} = (5.40 \pm 0.38) \cdot 10^6 \end{aligned} \quad (6.26)$$

with the error summed up $\triangle N = \sqrt{(\triangle N_{\gamma\gamma\text{-Analysis}})^2 + (\triangle N_{3\pi^0\text{-Analysis}})^2}$.

6.3.1. Branching Ratio of $\eta \to e^+e^-\gamma$

With the number of produced η-mesons in total the branching ratio of the η-Dalitz decay was calculated as:

$$\begin{aligned} BR_{\eta\text{-Dalitz}} &= \frac{\text{Counts}_{\text{Dalitz}} \cdot \frac{1}{\text{Acceptance}_{\text{Dalitz}}}}{N_{\text{Produced }\eta}} \\ &= \frac{436 \cdot \frac{1}{0.013}}{5.43 \cdot 10^6} \\ &= (6.177 \pm 0.649) \cdot 10^{-3} \end{aligned} \quad (6.27)$$

with the error determined in the following manner:

$$\triangle BR = \sqrt{|\frac{1}{NA}|^2 \cdot \triangle C^2 + |\frac{C}{N^2A}|^2 \cdot \triangle N^2 + |\frac{C}{NA^2}|^2 \cdot \triangle A^2} \quad (6.28)$$

$$\triangle BR = 0.000649 \quad (6.29)$$

This result corresponds to the PDG value, which is: $(6.0 \pm 0.8) \cdot 10^{-3}$. It has to be mentioned that in this work the reconstructed distribution of the invariant $m_{e^+e^-}$ starts at ~ 40 MeV (the same holds for the acceptance).

6.3.2. Branching Ratio of $\eta \to \pi^+\pi^-\pi^0$

In the same manner this was done for the η-decay into $\pi^+\pi^-\pi^0$:

$$\begin{aligned} BR_{\eta \to \pi^+\pi^-\pi^0} &= \frac{\text{Counts}_{\pi^+\pi^-\pi^0} \cdot \frac{1}{\text{Acceptance}_{\pi^+\pi^-\pi^0}}}{N_{\text{Produced }\eta}} \\ &= \frac{6956 \cdot \frac{1}{0.00056}}{5.43 \cdot 10^6} \\ &= 0.2287 \pm 0.017 \end{aligned} \quad (6.30)$$

With this error calculated according to equation 6.28. This result corresponds to the PDG value, which is $(22.7 \pm 0.4)\%$.

6.3.3. Branching Ratio of $\omega \to \pi^0\gamma$

In the analysis of $\omega \to \pi^+\pi^-\pi^0$ 8414 events were reconstructed and with the known branching ratio the number of ω-mesons produced in total was calculated as $1.576 \cdot 10^6$. Using this number as well as the acceptance of $\omega \to \pi^0\gamma$, which is 0.43% (section 6.1.6), the branching ratio of $\omega \to \pi^0\gamma$ was determined as (the error was determined according to equation 6.28):

$$\begin{aligned} BR_{\omega \to \pi^0\gamma} &= \frac{\text{Counts}_{\pi^0\gamma} \cdot \frac{1}{\text{Acceptance}_{\pi^0\gamma}}}{N_{\text{Produced }\omega}} \\ &= \frac{690 \cdot \frac{1}{0.0043}}{1.576 \cdot 10^6} \\ &= 0.1018 \pm 0.0135 \approx (10.2 \pm 1.4)\% \end{aligned} \quad (6.31)$$

This result is slightly higher than the corresponding PDG value: $8.9^{+0.27}_{-0.23}$. Still, within the errors the result agrees with the PDG value. The ratio of the two analyzed ω-decays was found to be[69]:

$$\frac{\Gamma_{\omega \to \pi^0\gamma}}{\Gamma_{\omega \to \pi^+\pi^-\pi^0}} = \frac{689.45/0.0043}{8360.3/0.00599} = 0.114 \pm 0.015 \quad (6.32)$$

$$\text{PDG: } \frac{\Gamma_{\omega \to \pi^0\gamma}}{\Gamma_{\omega \to \pi^+\pi^-\pi^0}} = 0.0998 \pm 0.003$$

6.4. Discussion of Background channels

It is not only of importance to remove the background accurately, but also to understand the origin of the background in the particular analyses. In this respect several Monte Carlo simulations were performed in order to identify possible background sources.

6.4.1. Background channels in the analysis of $\eta \to e^+e^-\gamma$

The reactions and decays of mesons that lead to the same final state definitely contribute to the background. Investigations of several simulated data were performed to determine possible background channels. Thereby it was found that for each investigated channel a certain fraction of started events entered in the final state $e^+e^-\gamma$ (proton). Hence, these events enter in the η-Dalitz analysis as background. However, the question was, how many of these events, and of which channel/reaction, survive all of the applied cuts? In the following the results of these investigations are presented for each background channel.

---
[69]The error was calculated according to equation 6.25.

6.4. Discussion of Background channels

Background from $\eta \to \gamma\gamma$

There are two mechanism that can lead to a contribution of $\eta \to \gamma\gamma$ to the η-Dalitz analysis. One is the so-called *conversion*. This effect describes the conversion of one photon into an e^+e^--pair. At first sight it seems not be so trivial to determine the correct percentage contribution of this effect. The reason for this is the second mentioned mechanism. It can happen that the a neutral particle is misidentified as a charged particle (electron/positron); and, of course, this can happen twice. Thus a $\gamma\gamma$-pair would be detected as an e^+e^--pair, which happens with a probability of approximately $2.2 \cdot 10^{-4}\%$[70], according to simulations for the developement of electromagnetic showers in the detector.

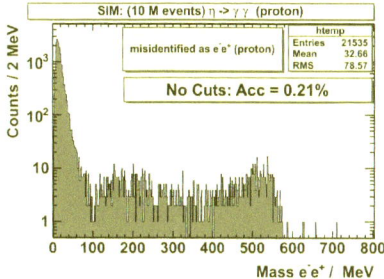 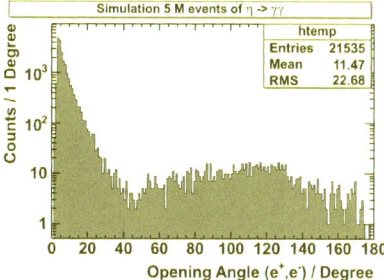

Figure 6.95.: Simulation of 10M events of $\eta \to \gamma\gamma$: plotted are the invariant masses of 2γ-pairs that were misidentified as e^+e^-.

Figure 6.96.: Simulation of 10M events of $\eta \to \gamma\gamma$: plotted is the opening angle of the detected e^+, e^- before cuts.

But, as a final state of $e^+e^-\gamma$ is required, an additional effect needs to take place in oder to make the misidentified e^+e^- contribute to the η-Dalitz analysis. This effect is the already mentioned split-off effect (section 6.1.3), which occurs with an probability of $\approx 0.13\%$[71]. Thus one can estimate[72] that an upper limit for the contribution of the misidentification effect is $2.86 \cdot 10^{-7}$, which is smaller than the branching ratio of the η-Dalitz decay ($BR \approx 6 \cdot 10^{-3}$). As roughly 5.5 million η-mesons were produced in the data[73], approximately 2.1 million decaying into $\gamma\gamma$, this effect does not play any role at all.

---

[70] The probability for a misidentification of a photon as $e^{\pm}$ is 1.48% (equation 6.36). This value was determined in the investigation of the background channel $\pi^0\gamma$ - page 185, 186.
[71] Determined in a simulation of η-Dalitz events using a cluster threshold of 50 MeV (section 6.1.3).
[72] Based only on the investigation of Monte Carlo simulations.
[73] This refers only to one beamtime (2007-07-LH_2).

6. Analysis

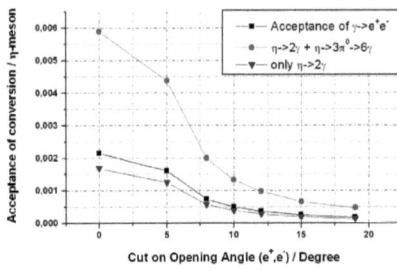

Figure 6.97.: Acceptance for the conversion process (black line) as a function of the cut on the e^+e^--opening angle. This acceptance was normalized to the number of produced photons per η-meson (see text).

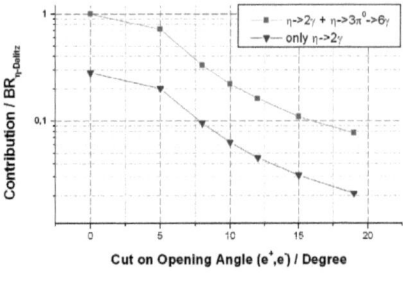

Figure 6.98.: Contribution in % of the conversion processes to the η-Dalitz channel as a function of the cut on the e^+e^--opening angle (see text).

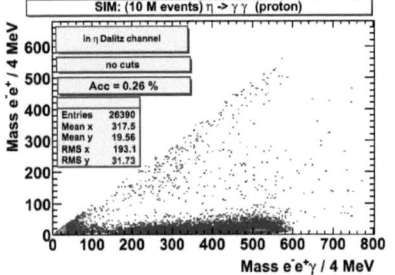

Figure 6.99.: Simulation of 10M events $\eta \to \gamma\gamma$: misidentified as $e^+e^-\gamma$ (before kinematic cuts).

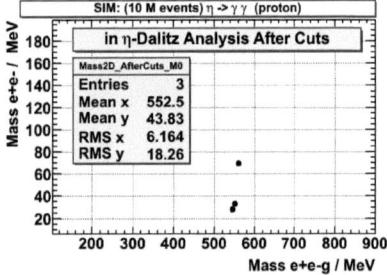

Figure 6.100.: Simulation of 10M events $\eta \to \gamma\gamma$: misidentified as $e^+e^-\gamma$. Only 3 counts survive the whole series of cuts on the kinematics.

Hence, it has to be concluded that nearly all simulated $\eta \to \gamma\gamma$-events detected as $e^+e^-\gamma$ are stemming from a conversion of one of the two photons. Thus the probability for detecting a conversion process is approximately 0.21%, as can be obtained from Figure 6.95. A cluster threshold of 50 MeV was used. If one applies a cluster threshold of only 5 MeV and a wider $dEvE$-banana cut as illustrated by in Figure A.16, this probability is increased to 1.3% (Figure A.15[74]). The conversion process was studied further. In

---

[74]In Figure A.16 the invariant e^+e^--mass of 85000 reconstructed conversion events is shown (detection of the recoiling proton was required). As 10 million events of $\eta \to \gamma\gamma$ were started in this simulation,

6.4. Discussion of Background channels

Figure 6.96 the opening angle of the detected e^+e^- is shown[75]. The number of entries in this histogram was integrated for several angluar ranges corresponding to different cuts on the opening angles (Table 6.25). The acceptance of e^+e^- as a function of the cut on the opening angle is illustrated by the black curve in Figure 6.97. The results were normalzied on the number of photons per produced η-meson, which is:

$$\text{for:} \quad \eta \to 2\gamma \text{ and } \eta \to 3\pi^0 \to 6\gamma$$
$$f_{N_1} = 2.7382\gamma = 2\gamma \cdot 0.3938 + 6\gamma \cdot 0.3251 \tag{6.33}$$

The normalized[76] values of the acceptance of $\gamma \to e^+e^-$ are listed in the 3rd and the fifth column of Table 6.25 (Normalized$_A$, Normalized$_B$). Thereafter the contribution of these normalized conversion processes to the η-Dalitz analysis was calculated by dividing by the branching ratio of the Dalitz decay. The result of this is listed in the columns four and six. The Figure 6.98 shows a plot of these contributions. If the contribution of the conversion effects shall be on the level of 2% a cut of $19°$ on the opening angle of e^+, e^- is required. It was found later, that the channel $\eta \to 3\pi^0$ did not contribute to the background after cuts (see: background from $\eta \to \pi^0\pi^0\pi^0$, Figure 6.106). In order to estimate an effective number (upper limit) of counts that contribute to the η-Dalitz analysis and that stemm from a convserion process, one has to multiply the number of produced η-mesons by the branching ratio of $\eta \to 2\gamma$, by the acceptance of the channel of interest[77] and by 0.021 (Table 6.25, last line, right colomn). The resulting value is 6 counts.

Angle-Cut in degree	Acceptance of $\gamma \to e^+e^-$	Normalized$_A$ f_{N_1}	Contribution η-Dalitz	Normalized$_B$ f_{N_2}	Contribution η-Dalitz
0	2.153e-3	5.89e-3	0.996	1.68e-3	0.28
5	1.597e-3	4.37e-3	0.72	1.24e-3	0.20
8	7.312e-4	2.0e-3	0.33	5.71e-4	0.095
10	4.858e-4	1.32e-3	0.22	3.78e-4	0.063
12	3.530e-4	9.66e-4	0.161	2.75e-4	0.045
15	2.394e-4	6.55e-4	0.109	1.86e-4	0.031
19	1.696e-4	4.64e-4	0.077	1.31e-4	0.021

Table 6.25.: Investigation of the contribution of the conversion effect to the background in the η-Dalitz analysis (see text).

In the analysis not only a cut on the opening angle is applied (Table 6.3). As is shown in

---

the probability for a (detected) conversion is $P_{\gamma\gamma \to e^+e^-} = \frac{85000}{10000000} \cdot 0.7 \approx 0.013$ (probability for the detection of a proton is $\sim 70\%$).

[75] For an simulation of $\eta \to 2\gamma$.

[76] In order to set the focus only on the contribution of $\eta \to \gamma\gamma$-events the normalization factor has to be calculated as: $f_{N_2} = 0.78\gamma = 2\gamma \cdot 0.3938$.

[77] Which is the η-Dalitz channel, that has an acceptance of 1.3 %.

6. Analysis

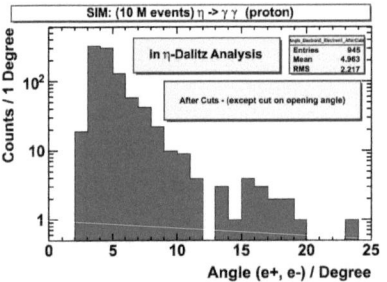

Figure 6.101.: Simulation of 10M events $\eta \rightarrow \gamma\gamma$: after a conversion of one γ into e^+e^-, the opening angle of the charged lepton pair is rather small.

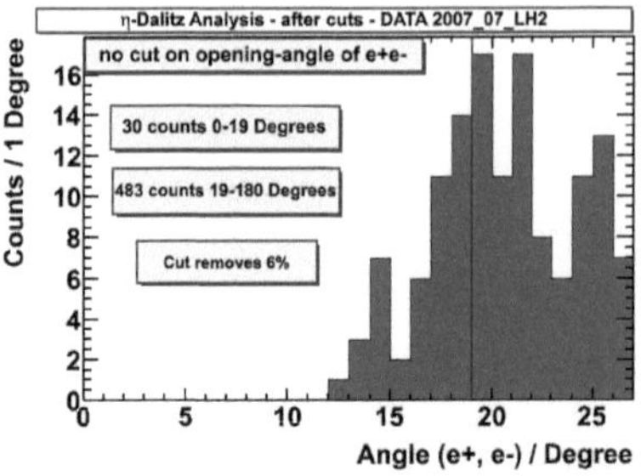

Figure 6.102.: Data: analysis of η-Dalitz events. A minimum opening angle of 19° is required.

Figure 6.99 only three counts survive the cuts applied in the η-Dalitz analysis. A strong reduction of events was realized by the cut on the opening angle of the electron/positron pair. Figure 6.101 shows this opening angle for the background events (after other cuts). As in the η-Dalitz a minimum opening angle of (e^+e^-) of 19° is required, nearly all background stemming from $\eta \rightarrow \gamma\gamma$ is removed.

Background from $\eta \rightarrow \pi^-\pi^+\gamma$

As the decay $\eta \rightarrow \pi^-\pi^+\gamma$ has a larger branching ratio than the η-Dalitz decay one might think, that this channel has to contribute in a strong manner to the final state of $e^+e^-\gamma$[78]. Clearly, this decay has to end up in the same final state as the η-Dalitz decay (before cuts), as in this experiment either e^+/e^-- or $\pi^+\pi^-$ can be reconstructed, due to the fact, that either a $\pi^{+/-}$-banana-cut or $e^{+/-}$-banana-cut can by used in the application of the dEvE method of particle identification. Figure 6.103 illustrates this fact.

Fortunately the background from charged pions can be suppressed very successfully, as it has been described in section 4.3.3. The usage of cuts on the missing mass, momentum balance and the energy balance (Figures 4.11 and 4.12) leads to an complete removal of these background events (Figure 6.104). Hence, the suppression factor of the $\pi^+\pi^-X$-channels[79] can be estimated as:

$$F_{suppress_{\pi^+\pi^-X}} = \frac{1}{3600000} \approx 2.77 \cdot 10^{-7} \quad (6.34)$$

---

[78]This experiment does not use a magnetic field. Hence, the separation of e^+e^- from $\pi^+\pi^-$ is not trivial.
[79]The X can either be a photon or an π^0.

6.4. Discussion of Background channels

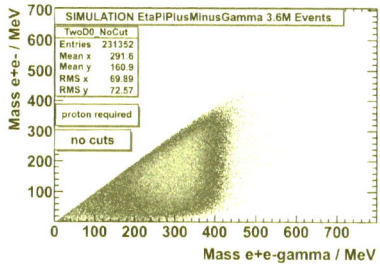

Figure 6.103.: Simulation of 3.6M events $\eta \to \pi^+\pi^-\gamma$: before the cuts are applied in the η-Dalitz analysis.

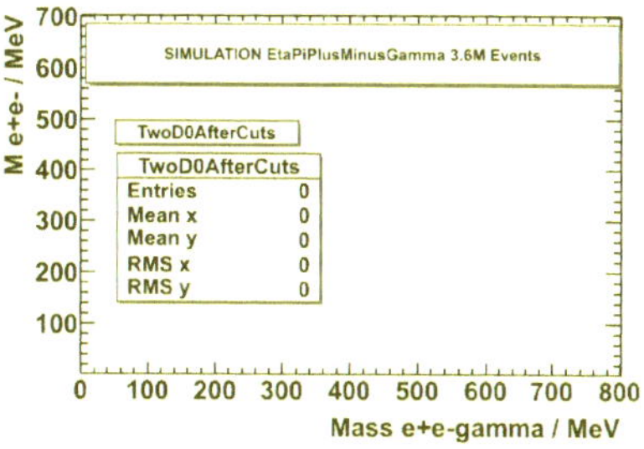

Figure 6.104.: Simulation of 3.6M events $\eta \to \pi^+\pi^-\gamma$: after the cuts are applied in the η-Dalitz analysis the 2D-plane is empty.

Background from $\eta \to \pi^-\pi^+\pi^0$

It has been shown in section 4.3.3., that the channel $\eta \to \pi^-\pi^+\pi^0$ does not contribute to the background.

Background from $\eta \to \pi^0\pi^0\pi^0$

One million events of $\eta \to \pi^0\pi^0\pi^0$ were simulated. As each of the pions can decay via the Dalitz channel or decays into two photons, one performing a conversion, it was assumed, that this channel might contribute to the η-Dalitz analysis. Fortunately all events were suppressed by the applied cuts (Table 6.3); this result can be obtained by a comparison of the two Figures 6.105 and 6.106. In the former Figure an indication for a misidentified[80] π^0-band is seen.

Background from $\eta \to \pi^0\gamma\gamma$

Although the decay $\eta \to \pi^0\gamma\gamma$ has a very small branching ratio[81], a possible contribution based on misidentification or conversion-processes was investigated. As three million events were simulated, the number of events entering the final state of the η-Dalitz

---
[80] $\pi^0 \to \gamma\gamma$ misidentified as $\pi^0 \to e^+e^-$.
[81] $BR_{\eta \to \pi^0\gamma\gamma} = 4.4 \cdot 10^{-4}$. The decay is even weaker than the η-Dalitz decay.

6. Analysis

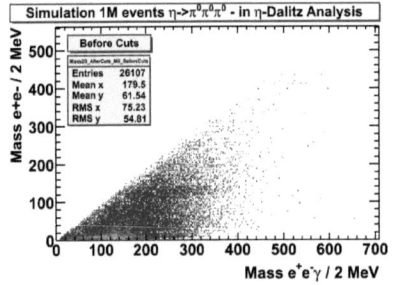
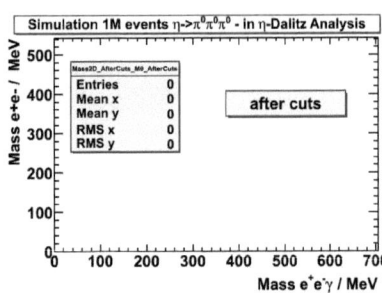

Figure 6.105.: Simulation of 1M events of $\eta \to \pi^0\pi^0\pi^0$: plot of $m_{e^+e^-}$ against $m_{e^+e^-\gamma}$ before cuts (see text).

Figure 6.106.: Simulation of 1M events of $\eta \to \pi^0\pi^0\pi^0$: plot of $m_{e^+e^-}$ against $m_{e^+e^-\gamma}$. No events survived the cuts.

analyses is far greater than expected in reality. However, some of the started events survive all the cuts applied in the analysis. This fact is illustrated by the Figures 6.107 and 6.108. In the former Figure a band of misidentified π^0-mesons can be seen at $m_{e^+e^-} \approx 135$ MeV. Still, this channel can not really contribute to the background because of its very small branching ratio.

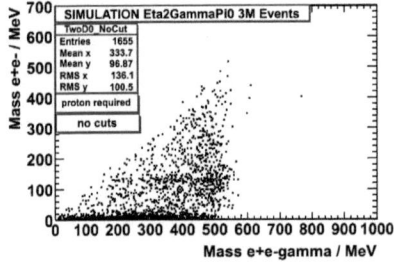

Figure 6.107.: Simulation of 3 M events of $\eta \to \pi^0\gamma\gamma$: plot of $m_{e^+e^-}$ against $m_{e^+e^-\gamma}$ before cuts (see text).

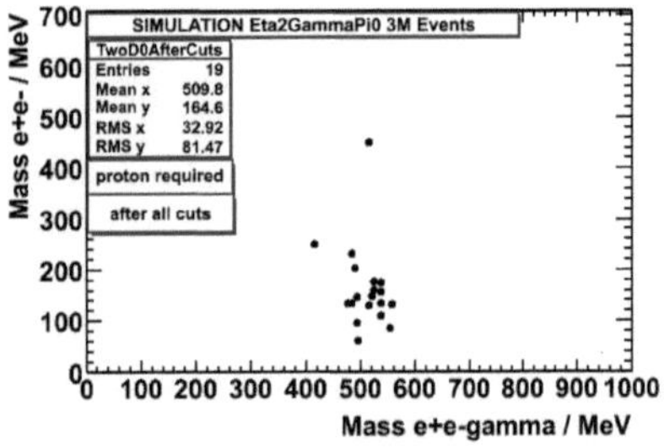

Figure 6.108.: Simulation of 3 M events of $\eta \to \pi^0\gamma\gamma$: plot of $m_{e^+e^-}$ against $m_{e^+e^-\gamma}$. Some events survived the cuts.

Background from $\omega \to e^+e^-\pi^0$

In the ω-Dalitz decay a π^0 is produced, which decays into two photons. If one of these photons is not detected, which happens with a probability of 12%, the same

6.4. Discussion of Background channels

final state is observed as in the case of a real η-Dalitz analysis. However, as higher multiplicities are analyzed all ω-Dalitz events have to contribute to the background in the η-Dalitz analysis. This fact is illustrated by the Figure 6.109 and 6.110. Even after cuts many events survive. The branching ratio[82] of the ω-Dalitz decay is, however, so small that even no ω-signal was detected in the investigation of the ω-Dalitz decay in the experimental data (section 6.2.8). Hence, this channel does not contribute to the background.

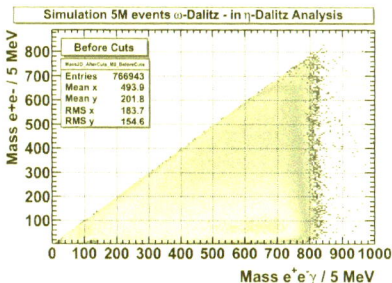

Figure 6.109.: Simulation of 5 M events of $\omega \to e^+e^-\pi^0$: plot of $m_{e^+e^-}$ against $m_{e^+e^-\gamma}$ (before cuts).

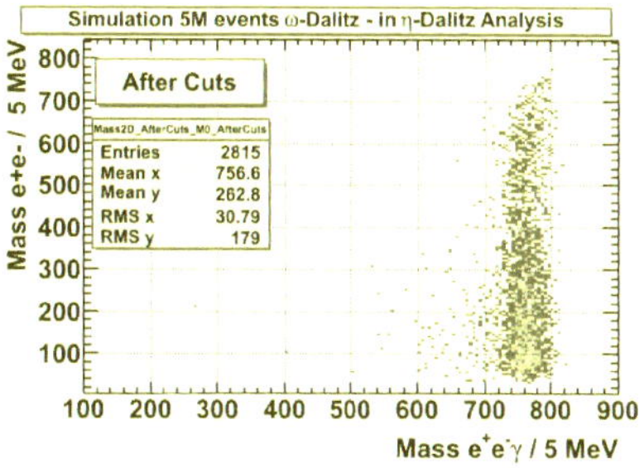

Figure 6.110.: Simulation of 5 M events of $\omega \to e^+e^-\pi^0$: plot of $m_{e^+e^-}$ against $m_{e^+e^-\gamma}$. Many events survived the cuts.

Background from $\omega \to \pi^+\pi^-\pi^0$

As with the corresponding decay of the η-meson, this channel does not contribute to the background in the final results of the η-Dalitz analysis (after cuts). The reason for this is again, that the misidentification of $\pi^+\pi^-$ as e^+e^- leads to different kinematics and thus these events can be removed by the application of the corresponding cuts. The results are shown in the Figure 6.111 and 6.112.

Background from $\omega \to \pi^0\gamma$

If one of the three photons[83] makes a conversion into an e^+e^--pair, the resulting final state enters in the η-Dalitz analysis because higher multiplicities are investigated. After the application of all cuts in the η-Dalitz analysis, still some counts survive. As in the analysis of the η-Dalitz decay a cut is applied on the incident energy (750 MeV to 1210 MeV) it can be assumed, that the number of produced ω-mesons for γ-beam energies

---
[82] $BR_{\omega \to e^+e^-\pi^0} = 7.7 \cdot 10^{-4}$, Particle Data Booklet 2006.
[83] Assuming the pion decayed into two photons before.

6. Analysis

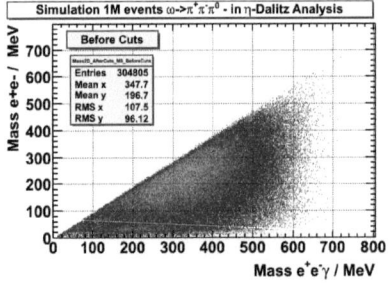 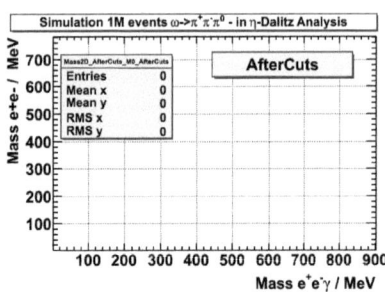

Figure 6.111.: Simulation of 1 M events of $\omega \to \pi^+\pi^-\pi^0$: plot of $m_{e^+e^-}$ against $m_{e^+e^-\gamma}$ (before cuts).

Figure 6.112.: Simulation of 1 M events of $\omega \to \pi^-\pi^-\pi^0$: plot of $m_{e^+e^-}$ against $m_{e^+e^-\gamma}$. No events survive the cuts.

up to 1210 MeV is far less than the number of produced η-mesons[84]. Furthermore the decay of $\omega \to \pi^0\gamma$ is not very strong (BR = 8.9 %). Thus there are far less events of this type in the data. Hence, this channel does not really contribute to the analysis of the η-Dalitz decay; and in case it would, the data points would not enter in the mass range of the η-meson (Figures 6.113 and 6.114).

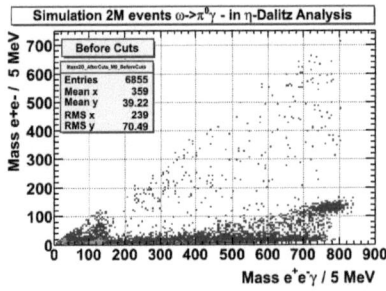 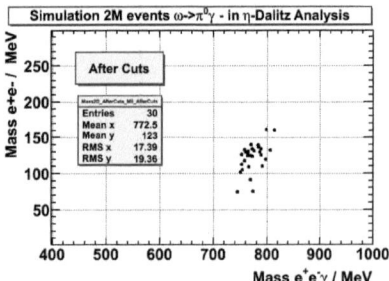

Figure 6.113.: Simulation of 2 M events of $\omega \to \pi^0\gamma$: plot of $m_{e^+e^-}$ against $m_{e^+e^-\gamma}$ (before cuts).

Figure 6.114.: Simulation of 2 M events of $\omega \to \pi^0\gamma$: plot of $m_{e^+e^-}$ against $m_{e^+e^-\gamma}$.

The analysis of the $\pi^0\gamma$-channel provides the possibility to investigate the misidentification process. Assuming both photons stemming from the decay of the π^0 (into $\gamma\gamma$) were misidentified as an e^+e^--pair, the reconstructed invariant e^+e^--mass has to be close

---

[84]Compare the measured production cross sections of the η and the ω-meson for a photon induced production off the proton [48].

6.4. Discussion of Background channels

to the π^0-mass (≈ 135 MeV). In Figure 6.115[85] several events of this type ($\pi^0 \to \gamma\gamma$ misidentified as $\pi^0 \to e^+e^-$) were identified. In Figure 6.116 the invariant e^+e^--mass is shown of the 438 events, that survived the 2D-cut (red curve) in Figure 6.115; As the number of started events in this simulation was 2 million, the probability for a misidentification of a $\gamma\gamma$-pair as an e^+e^--pair is given by (before cuts):

$$P_{\gamma\gamma \to e^+e^-} = \frac{438}{2000000} = 2.19 \cdot 10^{-4} \qquad (6.35)$$

Thus, the probability of a single misidentification is:

$$P_{\gamma \to e^{\pm}} = \sqrt{2.19 \cdot 10^{-4}} = 0.01478 \approx 1.5\% \qquad (6.36)$$

The two Figures A.12 and A.13 (appendix) show the opening angle of the misidentified e^+e^--pair and the opening angle of $e^{\pm}\gamma$. A comparison to the applied cuts in the η-Dalitz analysis (Table 6.3) shows, that these events can not be removed by cuts on these opening angles.

In the same manner the corresponding probability can be calculated after the cuts were applied; as can be seen from Figure 6.114 only 25 counts have to be considered as misidentified pions. Thus the probability is given by:

$$P_{\pi^0 \to \gamma\gamma \to e^+e^-} = \frac{25}{2000000} = 1.25 \cdot 10^{-5} \quad \text{(after cuts)} \qquad (6.37)$$

The reason for the misidentification to take place can either be, that the photon fired the PID/VETO, or that the photon passes through an PID/VETO-channel, which has been fired by an charged particle before[86]. Furthermore a electron or positron of the produced electromagnetic shower can travel in backwards direction and hit the VETO, which would be fired and as a result, the former neutral hit becomes marked as 'charged' (see: Background from $\pi^0\pi^0$).

Background from $\pi^0\eta$

The $\pi^0\eta$-production contributes also to the background in the η-Dalitz analysis. If the pion and the η decay both into two photons the final state of $e^+e^-\gamma$ can be generated if one of the four photons is not detected and the other one undergoes a conversion $\gamma \to e^+e^-$. Even if both photons were detected, such events still contribute, because higher multiplicities are analyzed.

Another possibility is, that the π^0 decays via a real Dalitz decay[87]; assuming the η-meson decays into two photons, this leads to the following final state: $e^+e^-\gamma\gamma$. As

---

[85]This is the same Figure as 6.113, but with a zoom on higher invariant masses.
[86]A PID/VETO-Channel that was fired, is marked as 'fired' until the time of the event-interval has passed (750 nsec).
[87]π^0-Dalitz decay: $\pi^0 \to e^+e^-\gamma$; $BR = 1.198\%$.

6. Analysis

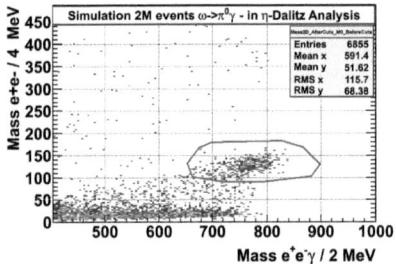

Figure 6.115.: Simulation of 2 M events of $\omega \to \pi^0\gamma$: plot of $m_{e^+e^-}$ against $m_{e^+e^-\gamma}$ (before cuts). The misidentified $\pi^0 \to \gamma\gamma$-events are marked by the red curve of the 2D-cut.

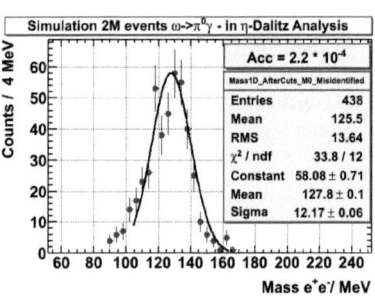

Figure 6.116.: Simulation of 2 M events of $\omega \to \pi^0\gamma$: reconstructed invariant e^+e^--mass of the misidentified events (see text).

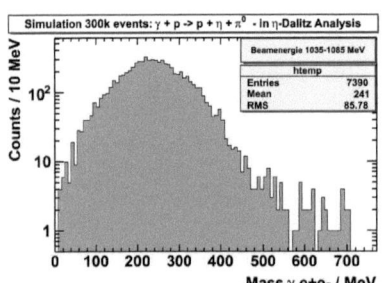

Figure 6.117.: Simulation of 0.3M $\pi^0\eta$-events: reconstructed invariant $e^+e^-\gamma$-mass of $\pi^0\eta$-production events.

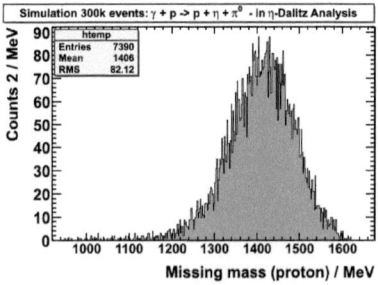

Figure 6.118.: Simulation of 0.3M $\pi^0\eta$-events: plot of the missing mass. Using an appropriate cut (Table 6.3) leads to a strong reduction of this background channel.

in the η-Dalitz higher multiplicities are taken into account, three combinations of these particles to $e^+e^-\gamma$ are possible.

Figure 6.117 shows the reconstructed invariant $e^+e^-\gamma$-mass after an investigation of 300.000 simulated events of $\pi^0\eta$-production. In Figure 6.118 the corresponding missing mass (proton) is plotted. Hence, this background can be reduced effectively by using cuts on the kinematics (for instance of the *missing mass*, which is shown in Figure 6.118). In Figure 6.119 the two dimensional plane of $m_{e^+e^-}$ versus $m_{e^+e^-\gamma}$ is shown before cuts;

6.4. Discussion of Background channels

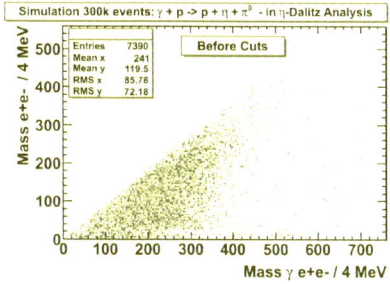

Figure 6.119.: Simulation of 0.3M $\pi^0\eta$-events: invariant mass of e^+e^- plotted versus $m_{e^+e^-\gamma}$ (before cuts).

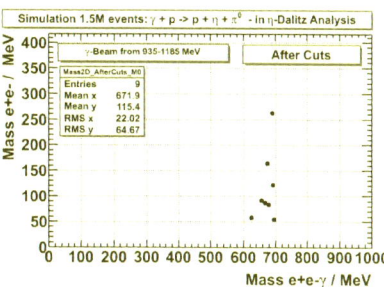

Figure 6.120.: Simulation of 1.5M $\pi^0\eta$-events: invariant mass of e^+e^- plotted versus $m_{e^+e^-\gamma}$ (after cuts).

Figure 6.120 shows the same plane after the application of all cuts and for an even more elaborate simulation of 1.5M events. The remaining background is not located in the regime of the η-mass.

Background from $\pi^0\pi^0$

For the same reason as for the previously discussed $\pi^0\eta$-production channel, the production of $\pi^0\pi^0$ also contributes to the background in the η-Dalitz analysis. Five million events were simulated and analyzed using the $e^+e^-\gamma$-analysis routines. Figure 6.121 shows the reconstructed invariant e^+e^--mass plotted against the $e^+e^-\gamma$-mass before the application of cuts. The same plot after cuts is shown in Figure 6.122. As can be learned from this Figure the $\pi^0\pi^0$-production channel is a strong background channel. Out of five million events, 27 events end up in the η-Dalitz analysis surviving all cuts. As the cross section for $\pi^0\pi^0$-production is comparable to the corresponding cross section for η-production it has to be assumed, that the number of $\pi^0\pi^0$-events in the data is comparable the number of η-mesons. Still, five million events in the simulation seem to be not enough. Unfortunately it was not possible to simulate more events because of limited computer time.

However, the $\pi^0\pi^0$-production was identified as the main source of background in the analysis of the η-Dalitz decay. The probability for a misidentification of a $\gamma\gamma$-pair as an e^+e^--pair is $2.19 \cdot 10^{-4}$ (equation 6.35 - before cuts). In the exclusive analysis of $\pi^0\pi^0$-events in the data (section 6.2.10) the upper limit of the probability for a pion to be misidentified and to contribute to the η-Dalitz analysis after cuts was determined as $1.7 \cdot 10^{-5}$. Figure 6.122 shows the result of a simulation of 5 million $\pi^0\pi^0$-events analyzed in the η-Dalitz analysis routine. As ≈ 20 counts are in the pion background band, the

6. Analysis

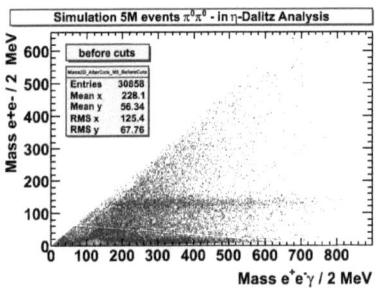

Figure 6.121.: Simulation of $\pi^0\pi^0$ in the η-Dalitz analysis. 2D-plot of $m_{e^+e^-}$ against $m_{e^-e^+\gamma}$ of the misidentified events (before cuts).

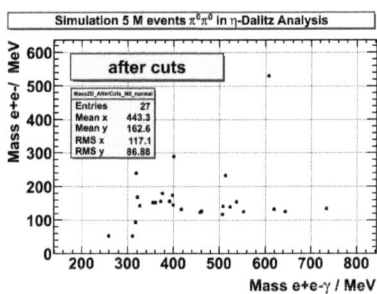

Figure 6.122.: Simulation of $\pi^0\pi^0$ in the η-Dalitz analysis. 2D-plot of $m_{e^+e^-}$ against $m_{e^-e^+\gamma}$ of the misidentified events (before cuts).

probability for a pion to be misidentified and to contribute to the η-Dalitz analysis after cuts has to be calculated as: $4 \cdot 10^{-6}$, which is 4 times smaller than the determined upper limit. If one adds the result in equation 6.37 ($P = 1.25 \cdot 10^{-5}$) the resulting probability (of both channels - after cuts) is $P = 1.69 \cdot 10^{-5}$, which is consistent with the upper limit determined in the data ($P = 1.7 \cdot 10^{-5}$, section 6.2.10).

Figure A.23 shows a projection of the 2D-plot in Figure 6.121 onto the $m_{e^+e^-}$-axis with a fit to the (misidentified) pion signal; the number of counts in the π^0-signal was determined as ≈ 2000. Thus the corresponding probability (before cuts) is given by:

$$P_{\pi^0 \to \gamma\gamma \to e^+e^-} = \frac{2000}{5000000} = 4 \cdot 10^{-5}$$

This result is comparable to the probability found in equation 6.35.

As in each $\pi^0\pi^0$-event four photons are produced[88] the probability of a misidentification per event is increased by a factor of 2 compared of the situation we have had in the case of the reaction $\gamma + p \to \eta + p \to \gamma\gamma + p$. This explains the strong band of misidentified pions in Figure 6.121.

The question was, hot to suppress these events. As far as the misidentification of $\gamma\gamma$ as e^+e^- is concerned, the assumption was made that most of the mis identifications happen in TAPS. As the VETO detectors are directly in front of the BaF_2-crystals it can happen, that an electron generated by the electromagnetic shower caused by the γ-hit, goes backwards out of the crystal into the VETO detector, which is fired. This assumption was proven correct by the histogram plotting the θ-angles[89] of all photons that were misidentified as electrons (Figure 6.123). Consequently it was tested, whether some of

---
[88] Assuming both pions decay into 2 photons.
[89] The angular range of θ that is covered by TAPS is $\pm 20°$.

6.4. Discussion of Background channels

these background events surviving the cuts, could be suppressed by applying an extra cut on the θ-angles of the e^+/e^- requiring at least a value of $22°$ (Figure 6.124). It was found, that this additional cut removed more signal-events but no background-events and thus it was not used in the accomplished η-Dalitz analysis. As a matter of fact, most misidentifications happen in TAPS; but the misidentified e^+e^- of those background events, that survive the cuts applied in the η-Dalitz analysis, are not detected in TAPS (Figure A.14 appendix).

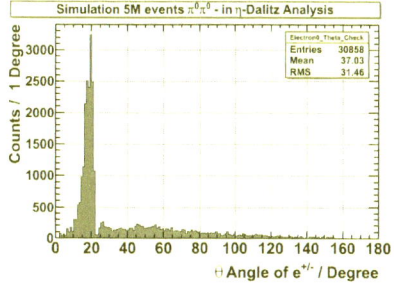

Figure 6.123.: Simulation of $\pi^0\pi^0$ in the η-Dalitz analysis. 1D-plot of the θ-angle of electrons in the η-Dalitz final state (before cuts).

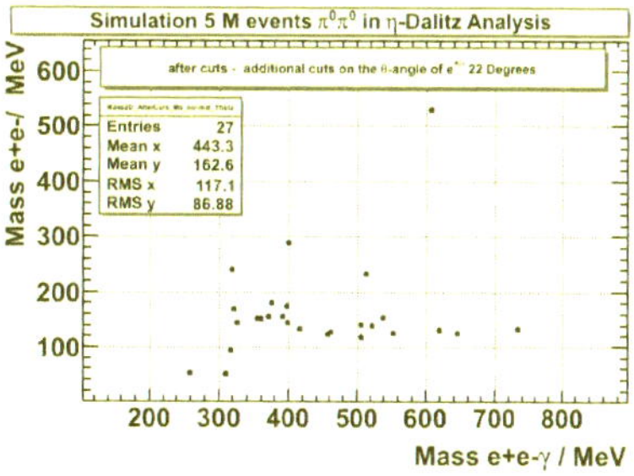

Figure 6.124.: Simulation of $\pi^0\pi^0$ in the η-Dalitz analysis. 2D-plot of $m_{e^+e^-}$ against $m_{e^-e^+\gamma}$ of the misidentified events (after cuts plus extra cut on the θ-angle of e^+e^-).

Another test in order to reduce the $\pi^0\pi^0$-background was to allow only incident photon energies less than 1 GeV, because the η-production cross section drops strongly above 1 GeV, while the $\pi^0\pi^0$-production cross section stays almost constant. It was shown by a simulation that this additional requirement reduced the $\pi^0\pi^0$-background by $\approx 10\%$ (Figure 6.125). On the other hand this requirment led to a loss of the same order of magnitude in the number of reconstructed η-Dalitz events. Hence, it was tested to compensate this loss by applying less strict cuts on the momentum balance. In this case the background contribution from $\pi^0\pi^0$ was increased again to a value larger than the original one (Figure 6.126).

Background from $\pi^+\pi^-$

As with all $\pi^{+/-}$-background events, the events from the production of $\pi^+\pi^-$ do not survive the cuts. Despite the strong cross section and the fact, that a simple split-off causes such events to contribute to the η-Dalitz analysis, all events can be removed by the cuts on the kinematics. The Figures 6.127 and 6.128 illustrate this fact.

6. Analysis

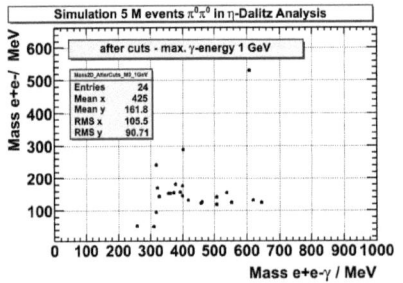

Figure 6.125.: Simulation of $\pi^0\pi^0$ in the η-Dalitz analysis. 2D-plot of $m_{e^+e^-}$ against $m_{e^-e^+\gamma}$ of the misidentified events (after cuts; with a maximum γ-beam energy of 1 GeV).

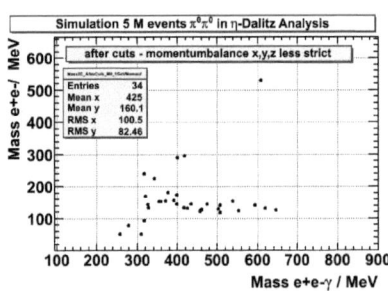

Figure 6.126.: Simulation of $\pi^0\pi^0$ in the η-Dalitz analysis. 2D-plot of $m_{e^+e^-}$ against $m_{e^-e^+\gamma}$ of the misidentified events (after cuts; with a maximum γ-beam energy of 1 GeV; all cuts on the momentum balance were 25% less strict).

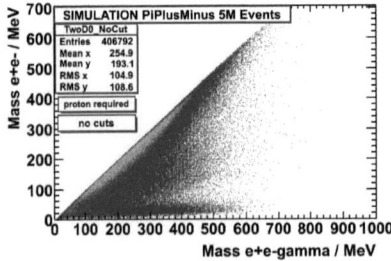

Figure 6.127.: Simulation of $\pi^-\pi^+$ in the η-Dalitz analysis.

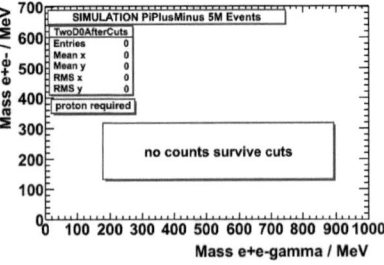

Figure 6.128.: Simulation of $\pi^+\pi^-$ in the η-Dalitz analysis. 2D-plot of $m_{e^+e^-}$ against $m_{e^-e^+\gamma}$ (after cuts). No events survive.

6.4.2. Background channels in the analysis of $\omega \to \pi^0 \gamma$

It was already mentioned, that the investigation of $\omega \to \pi^0\gamma$ is one of the main issues of the research work pursued by the A2 and CB/ELSA group of the University of Giessen. For all analyses it is always helpful to identify certain background channels. In case of this ω-decay every non-ω-event containing three neutral hits[90] contributes to the

---

[90] Neutral hits are always reconstructed as photons (section 4.3.1 and 4.3.2).

background. In section 6.2.9 and 6.2.10 the possible contribution of $\pi^0\pi^0$ and $\pi^0\eta$ was discussed. As these mesons decay, beyond other decay modes, into $\gamma\gamma$, a final state of three photons can easily occur, if one of the photons is not detected, which happens with a probability of $\approx 12\%$. Moreover, these events still enter in the analysis, even if no photon is missed. The reason for this is that higher multiplicities are taken into account. The background contribution from these two reactions was investigated and is shown in Figure 6.90.

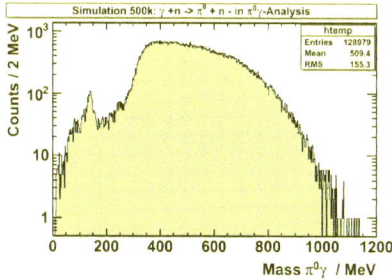
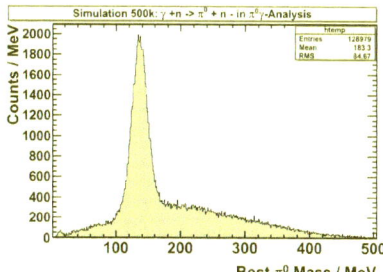

Figure 6.129.: Simulation: background in the analysis of $\omega \to \pi^0\gamma$. The production of π^0 off the neutron enters in the same final state, when the neutron is misidentified as γ.

Figure 6.130.: Simulation: π^0-production events (off the neutron) fulfill the mass cut on the best π^0, which is applied in the ω-analysis.

Another strong background source is the π^0-production off the neutron, which only contributes to the background in case of an inclusive analysis[91]. As members of the A2 and the CB/ESLA group investigated the decay $\omega \to \pi^0\gamma$ inclusively, this special background source had to be investigated[92]. A simulation of 500.000 events of $\gamma + n \to \pi^0 + n$ was performed and thereafter the output of the MC-simulation was analyzed using an inclusive 3γ-analysis function of $AR_{HB}2v3$. As with the corresponding exclusive function the combination of two photons out of three giving the *best* pion was requested using a χ^2-test. The χ^2-test is assumed to be always fulfilled, because a π^0 was really produced. If the recoiling neutron is misidentified as a photon this background will enter in the $\omega \to \pi^0\gamma$ analysis and will survive every cut on the mass of the *best* pion (Figures 6.129 and 6.130). Strict cuts on the timing spectra might help to suppress these events

---

[91] In an inclusive analysis the detection of the backscattered particle is not required (proton, neutron). On the one hand this leads to greater statistics, but on the other hand certain tests of the kinematic can not be applied (e.g. energy balance, momentum balance, etc.).

[92] It has to be mentioned, that this background has no connection to the analysis of $\omega \to \pi^0\gamma$ performed by the author, because of the usage of a proton target and the fact, that in the analysis a backscattered proton was required.

6. Analysis

because the slower neutrons will always hit the detectors with some delay (compared to the photons). As the time resolution of the Crystal Ball is only in the few ns range, and as the Crystal Barrel dos not provided any timing at all, this method is clearly limited. A means that really can help to reduce this background is to require a cut on

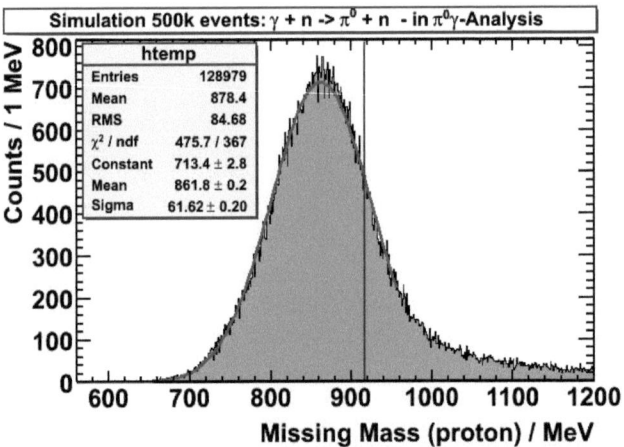

Figure 6.131.: Simulation of $\pi^0 + n$: plot of the missing mass in a 3γ-analysis.

missing mass (proton): $915 < mm < 965$. In this case a huge fraction of π^0-neutron events can be removed. Figure 6.131 shows the missing mass spectrum for a simulated π^0-production off the neutron in the analysis of a 3γ final state. Unfortunately this is not applicable in case of nuclear reactions on heavy targets because of the Fermi motion.

7. Results

In this chapter the results of this thesis are presented and discussed. The data were taken during one beamtime in July 2007 using a liquid hydrogen target. For the analysis of the η-Dalitz decay a second beamtime (June 2007, same target) was investigated in addition.

7.1. The measured channels

Several decay channels of the η-meson and the ω-meson were investigated. The transition form factor of the η-meson was determined in the analysis of the η-Dalitz decay (section 7.3). Further on the branching ratios of $\eta \to e^+e^-\gamma$, $\eta \to \pi^+\pi^-\pi^0$ and $\omega \to \pi^0\gamma$ were determined (section 6.3). The obtained results are:

$$BR_{\eta\text{-Dalitz}} = (6.177 \pm 0.649) \cdot 10^{-3}$$
$$BR_{\eta \to \pi^+\pi^-\pi^0} = (22.9 \pm 1.17)\%$$
$$BR_{\omega \to \pi^0\gamma} = (10.2 \pm 1.35)\%$$

These results are consistent with the values in the Particle Data Booklet of the Particle Data Group[1], which are: $BR_{\omega \to \pi^0\gamma} = 8.9^{+0.27}_{-0.23}\%$, $BR_{\eta \to \pi^+\pi^-\pi^0} = (22.7 \pm 0.4)\%$ and $BR_{\eta\text{-Dalitz}} = (0.6 \pm 0.08)\%$. The error obtained for the branching ratio of the η-Dalitz decay in this work is smaller than the corresponding error listed in the Particle Data Booklet.

Moreover the cross section of $\pi^0\eta$-production was measured (section 6.2.9). As can be seen from Figure 7.1 the result of this work is in agreement with the result obtained by [26], which has been published recently. The data points are listed in Table 6.22.

Furthermore the Dalitz decay of the π^0-meson was investigated (section 7.4).

Table 7.1 lists the results obtained in all analyses of this thesis. In the analysis of the η-meson the range of the incident γ-beam energy was 750 MeV to 1210 MeV. In case of the ω-meson the corresponding range was 1125 MeV to 1408 MeV; and in the analysis of the π^0-Dalitz decay the range was 610 MeV to 1410 MeV.

---

[1]Particle Data Booklet 2006.

7. Results

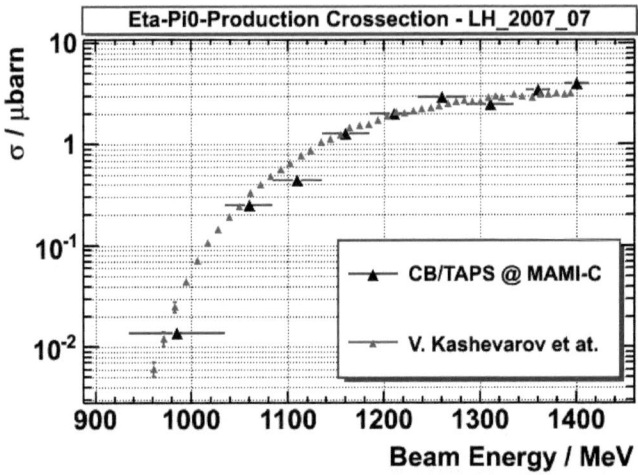

Figure 7.1.: Measured total cross section of $\pi^0\eta$-production in photon induced reactions off the proton (for incident energies up to 1408 MeV) in comparison to the result obtained in [26].

Decay	Counts	Acceptance [%]	Results
$\eta \to \gamma\gamma$	249500 ± 499	12.1	$N_{\eta_{produced}} = (5.23 \pm 0.26) \cdot 10^6$
$\eta \to \pi^0\pi^0\pi^0$	49855 ± 223	2.75	$N_{\eta_{produced}} = (5.57 \pm 0.28) \cdot 10^6$
$\eta \to \pi^+\pi^-\pi^0$	6956 ± 39	0.56	$BR = (22.9 \pm 1.7)\%$
$\eta \to e^+e^-\gamma$	436 ± 31	1.3	$b = 1.84^{+0.43}_{-0.32}$ GeV^{-2}
			$BR = (6.18 \pm 0.65) \cdot 10^{-3}$
$\eta \to \pi^0\gamma\gamma$	30 ± 19	–	–
$\omega \to \pi^0\pi^+\pi^-$	8414 ± 138	0.6	$N_{\omega_{produced}} = (1.58 \pm 0.08) \cdot 10^6$
$\omega \to \pi^0\gamma$	690 ± 77	0.43	$BR = (10.2 \pm 1.4)\%$
$\omega \to e^+ + e^-\pi^0$	–	1.08	–
$\pi^0 \to e^+ + e^-\gamma$	555 ± 27	0.2	$b_{\pi^0} \approx 1$

Table 7.1.: Listed are the analyzed decay channels, number of reconstructed events, the acceptance (for an exclusive analysis) and the results. The listed information only refer to the data set obtained in the July beamtime 2007.

7.2. The separation of e^+e^- from $\pi^+\pi^-$

As the CB/TAPS experiment does not use a magnetic field, a separation of e^+, e^- from π^+, π^- seems to be difficult. Nevertheless, this separation was achieved in this work[2]. The Crystal Ball as well as the TAPS detector use the dE-versus-E method in order to identify charged particles. Although this technique works fine in the case of protons[3], it is not an appropriate means to separate e^+, e^- from π^+, π^-, because these particles share the same regime in the 2D-plot of dE versus E (sections 4.3.1, 4.3.2). As only one banana-cut can be used in this regime, these particles are either reconstructed as e^+, e^- OR π^+, π^-. Thus, in case of a misidentification, the wrong particle-ID and the wrong mass are assigned to the detected particle. This of course affects the Lorentz vector of the reconstructed particle which can be exploited, as will be explained in the following.

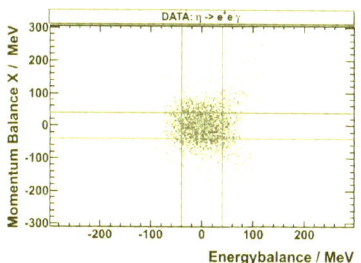

Figure 7.2.: Data: analysis of $\eta \rightarrow e^+e^-\gamma$ events. 2D-Plot of the *momentum balance in X* versus the *energy balance*.

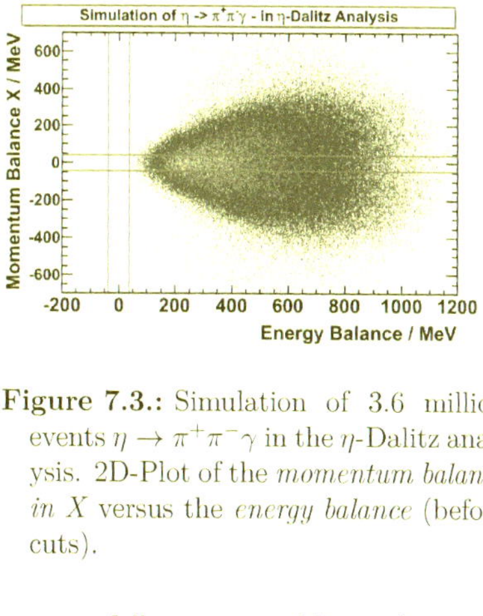

Figure 7.3.: Simulation of 3.6 million events $\eta \rightarrow \pi^+\pi^-\gamma$ in the η-Dalitz analysis. 2D-Plot of the *momentum balance in X* versus the *energy balance* (before cuts).

It was found, that the $\pi^\pm$-background can be successfully suppressed by exploiting the full kinematic information in an exclusive analysis. When the recoiling proton is detected[4] cuts on the momentum balance and the energy balance become applicable. Figures 7.2 and 7.3 illustrate how the $\pi^\pm$-background is suppressed by these cuts. In Figure 7.2 simulated η-Dalitz events were analyzed. If one uses the same analysis function in case of $\pi^+\pi^-\gamma$ events[5] (Figure 7.3), the cuts on the momentum balance and the energy balance (red lines) are not fulfilled.

---

[2]It has to be mentioned, that in this experiment e^+ and e^- are treated equally (because the sign of the charge can not be determined without a magnetic field). The same holds for π^+ and π^-.
[3]At least in the energy regime of this experiment.
[4]In an exclusive analysis the recoiling proton is detected.
[5]The $\pi^\pm$ are misidentified as $e^\pm$.

7. Results

Furthermore it was found, that cuts on missing mass and the cluster sizes of the charged hits further suppress the $\pi^{\pm}$-background[6] (section 4.3.3). The suppression factor was determined in section 6.4.1 (equation 6.34) as $F_{suppress_{\pi^+\pi^-X}} \approx 3 \cdot 10^{-7}$.

Hence, it has been shown in this thesis that a separation of e^+e^- from $\pi^+\pi^-$ can be realized without the use of a magnetic field.

7.3. The η-Dalitz decay

In the exclusive analysis of the η-Dalitz decay 827 events were reconstructed under the application of the cuts listed in Table 6.3 (section 6.2.4). These cuts represent an appropriate compromise between strictness and statistics. The stricter the cuts, the less events survive the cuts (this holds also for real η-events). On the one hand the signal to background ratio is improved by stricter cuts, on the other hand the statistics is decreased at the same time, which leads to difficulties in the measurement of the distribution of $m_{e^+e^-}$ for large invariant masses, which are close to the η-mass. In contrast: if one uses cuts that are not strict enough, the resulting signal to background ratio becomes inappropriate. Furthermore this leads to difficulties in the identification of real η-Dalitz events (especially for larger invariant masses of e^+e^-).

Table 6.3 presents the cut-setting found to be the most appropriate for the exclusive analysis of $\eta \to e^+e^-\gamma$ with CB/TAPS @ MAMI-C. These cuts were also applied in the investigation of the simulated data; the Monte-Carlo simulation was used to determine the correct detector response. The mass integrated overall efficiency was determined to be $\epsilon_{Dalitz} = 1.3\%$ for 750 MeV $\leq E_\gamma \leq 1210$ MeV (section 6.1.3). Figure 7.4 shows the distribution of the invariant $m_{e^+e^-}$ for reconstructed events after acceptance correction.

Due to the fact, that the scalers in the data of one run period were broken, only the data of the beamtime 2007-July-lH_2 can be used in the determination of the total cross section for η-production. Using only data of this beamtime 436 ± 31 η-Dalitz decays were identified. As in the analysis of the η-Dalitz decay a cut was applied on the energy of the incident γ-beam (750 MeV to 1210 MeV) the same interval had to be used in the determination of the corresponding photon flux (section 3.1.3). Hence, using the branching ratio determined in section 6.3.1 ($\Gamma_{Dalitz} = (6.2 \pm 0.6) \cdot 10^{-3}$), the photon flux ($N_{e^-} \cdot \epsilon_{tag}$, section 3.13) as well as the number of target atoms N_T the resulting total cross section was calculated as:

$$\sigma = \frac{N_{exp}}{N_{e^-} \cdot \epsilon_{tag} \cdot N_T \cdot \epsilon_{Dalitz} \cdot \Gamma_{Dalitz}}$$
$$\sigma_{\eta_{total}} = (5.01 \pm 0.92) \; \mu\text{barn} \qquad (7.1)$$

---
[6]Charged pions produce smaller clusters in the calorimeters then electrons/positrons (on average).

7.3. The η-Dalitz decay

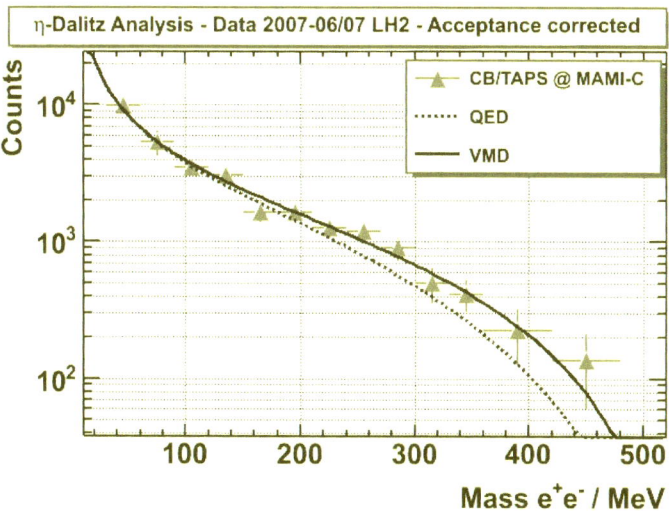

Figure 7.4.: Analysis of the η-Dalitz decay; distribution of $m_{e^+e^-}$ of the reconstructed events after acceptance correction. The dotted curve is the QED prediction scaled to the data points below 120 MeV. The solid curve is a fit within the VMD-model using a monopole form factor.

thereby the target density was determined in the following way:

$$N_T = \frac{N_A \cdot \rho_{target} \cdot L_{target}}{A}$$
$$= 2.02 \cdot 10^{-7} \frac{1}{\mu\text{barn}}$$

The error of the cross section was calculated in the following manner:

$$\triangle\sigma = \sqrt{|\frac{\triangle N_{exp}}{N_{e^-} \cdot \epsilon_{tag} \cdot N_T \cdot \epsilon_{Dalitz} \cdot \Gamma_{Dalitz}}|^2 + |\frac{N_{exp} \cdot \triangle\Gamma_{Dalitz}}{N_{e^-} \cdot \epsilon_{tag} \cdot N_T \cdot \epsilon_{Dalitz} \cdot \Gamma_{Dalitz}^2}|^2}$$
$$\cdot \sqrt{|\frac{\triangle\epsilon_{Dalitz} \cdot N_{exp}}{N_{e^-} \cdot \epsilon_{tag} \cdot N_T \cdot \epsilon_{Dalitz}^2 \cdot \Gamma_{Dalitz}}|^2}$$

This result is not in perfect agreement with a weighted estimation for σ based on data points that were provided by [48]: $\sigma = (7.4 \pm 0.5)$ μbarn. The reason for the slight discrepancy might be the uncertainty in the estimation of the dead time correction factor in the determination of the photon flux. Furthermore the cuts on the cluster sizes

7. Results

of the charged hits (e^+, e^-) remove more events generated by low energetic photons of the incident γ-beam than events produced by high energetic incident photons (the cross section is larger for lower incident photon energies).

However, the number of reconstructed η-Dalitz events is in a correct relation to the counts detected in the other investigated decay channels of the η-meson ($\eta \to \gamma\gamma$, $\eta \to \pi^0\pi^0\pi^0$, $\eta \to \pi^0\pi^-\pi^+$). After the application of the corresponding acceptance corrections, the determined numbers of produced η-Mesons agree for all of these decay channels (see equations: 6.7, 6.9, 6.12, and 6.15). In total $(5.43 \pm 0.32) \cdot 10^6$ η-mesons were produced in the run period in July 2007 (equation 6.26).

As a matter of fact, it was found that the number of reconstructed η-Dalitz decays in the other beamtime (2007-June-lH_2) is less, although the amount of taken data is larger (the experimental settings were the same[7]). Moreover the number of produced η-mesons per data file was not stable. Hence, this beamtime was only taken into account in the determination of the η-form factor, because in this respect all that mattered was statistics.

7.3.1. The transition form factor

The transition form factor describes the difference to the QED prediction. The form factor was determined by dividing of the data points in Figure 7.4 by the value of the integral of the QED-curve within the corresponding interval of $m_{e^+e^-}$. Thereafter the data points in the resulting histogram were fitted (Figures 7.5 and 7.6). The data points are listed in Table A.3 (appendix). In this thesis the slope of the transition form factor of the η-meson was determined as:

$$b_\eta = \frac{dF}{dq^2}\Big|_{q^2=0} = \frac{1}{\Lambda^2} = 1.84^{+0.43}_{-0.32} \frac{1}{GeV^2} \qquad (7.2)$$

with a fit parameter (section 6.2.4):

$$\Lambda = (740 \pm 74) \text{ MeV}$$

The results of former experiments (Table 1.6, first chapter) are in agreement with our result. The determined value of the slope parameter of the transition form factor in this work is also consistent with the values found by the NA60 and the Lepton-G experiment; in the former the decay $\eta \to \mu^+\mu^-(\gamma)$ was investigated and in the latter $\eta \to \mu^+\mu^-\gamma$ was analyzed.

An investigation of $\eta \to e^+e^-\gamma$, that has to be mentioned, was accomplished by M.

---
[7]The same trigger settings, beam energy as well as beam current were used.

N. Achasov et al. [1] in 2001. The results of [1] are shown in Figure 7.5. In this experiment 109 Dalitz decays were reconstructed and the slope of the form factor was determined as (1.6 ± 2.0) GeV^{-2}. Until today this published result of the η-form factor has been the result with the highest statistics[8]. As in the present work 827 η-Dalitz events were successfully reconstructed, the statistics is larger by a factor of 7.6. Thus, the result of this thesis is the most accurate value of the η-transition form factor derived from an investigation of the decay mode $\eta \to e^+e^-\gamma$.

In Figure 7.6 a comparison to the results of the NA60 experiment is shown. In contrast to the present work, the NA60 collaboration investigated the channel $\eta \to mu^+\mu^-(\gamma)$ inclusively, which means, that only the charged leptons were detected. The statistics in this experiment is 9000 counts and the obtained slope parameter of the η-transition form factor is: $b_\eta = (1.95 \pm 0.17 \pm 0.05)$ GeV^{-2}. Within the errors the result from NA60 agrees with the result of the present work.

Moreover, Figures 7.5 and 7.6 shows a fit to the experimental results published by L.G. Landsberg [31] (black line) and a model calculation by C. Terschluesen and S. Leupold [46] (green line). The data points obtained in the present work agree with this recent calculation ($b_{\eta_\text{theo}} = 1.79$ GeV^{-2}, pole parameter of $\Lambda = 747$ MeV). In Figure 7.7 a comparison between the result of this work and the results of NA60, Lepton-G and SND is shown.

7.4. The Dalitz decays of the π^0-meson

In the investigation of the π^0-Dalitz decay ($\pi^0 \to \gamma\gamma^* \to e^+e^-\gamma$) 555 ± 27 events were reconstructed (section 6.2.11). The background was removed using the side band subtraction method as in the analysis of the η-Dalitz decay.

Figure 7.8 shows the final result after acceptance correction. The black line illustrates the QED-prediction. The result of the present work (red data points) is in agreement with a published result by N.P. Samios [41] (blue points). As can be seen in this figure, the measured distribution of the invariant e^+e^--mass does not differ from the QED prediction; thus the form factor is ≈ 1.0. As a matter of fact, this result has been expected since the mass of the π^0-meson is ≈ 135 MeV and thus the coupling of π^0 to a γ^* via a vector meson[9] is suppressed. Hence, this expectation was confirmed by the result of this thesis. The data points are listed in Table 6.23.

---

[8]This refers only to the channel: $\eta \to e^+e^-\gamma$.
[9]The lightest vector meson is the ρ-meson, which has a mass of 770 MeV [20].

7. Results

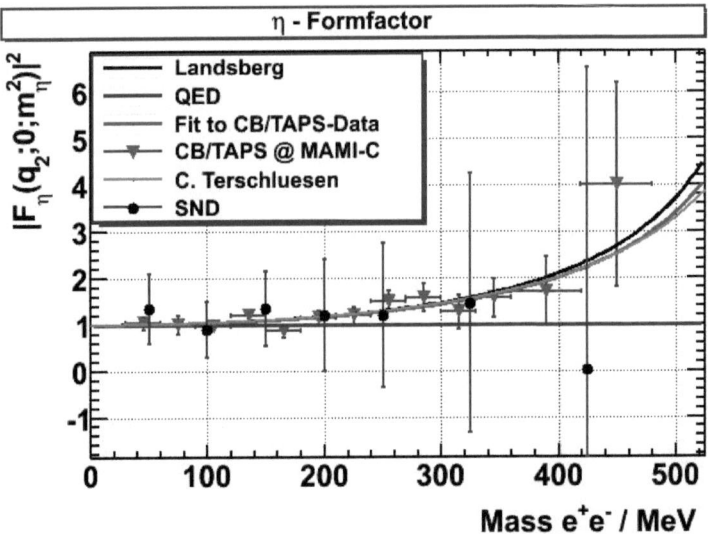

Figure 7.5.: Measurement of the η-Dalitz transition form factor. The red triangles are the data points of this work (the red line is the fit to the data). The black squares show the result of the SND experiment [1]. The green line shows a calculation performed by Terschluesen and Leupold. The black curve is the fit curve to the data of [31].

7.4. The Dalitz decays of the π^0-meson

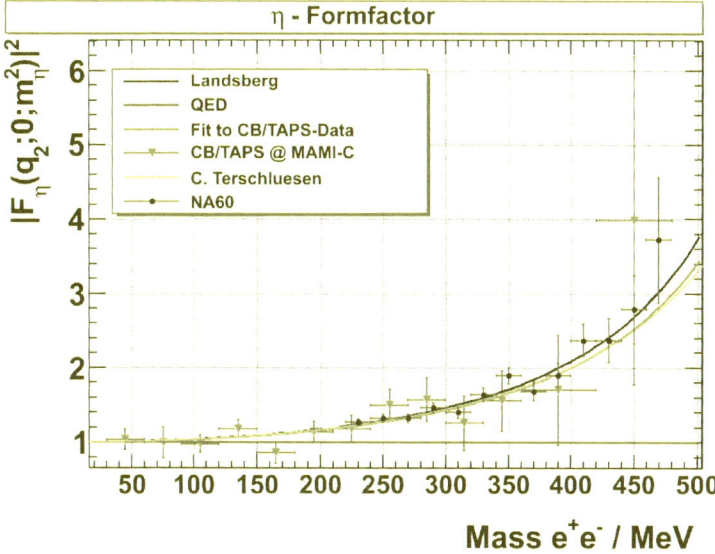

Figure 7.6.: Measurement of the η-Dalitz transition form factor. The red triangles are the data points (the red line is the fit to the data). The black points are the result from [12]. The green line shows a calculation performed by Terschluesen and Leupold. The black curve is the fit curve to the data of [31].

7. Results

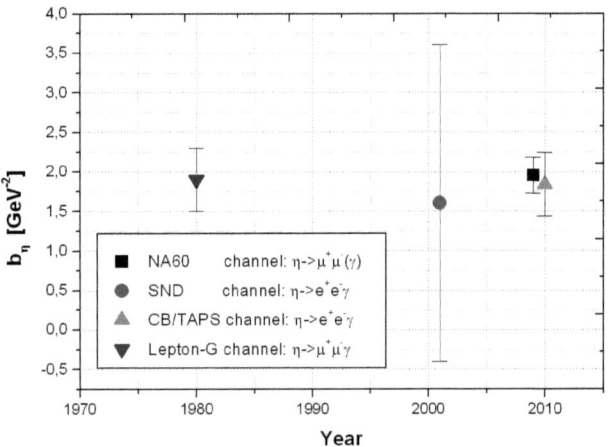

Figure 7.7.: The slope b_η of the η transition form factor. The result of this work is shown in comparison to former results (SND, NA60, Lepton-G).

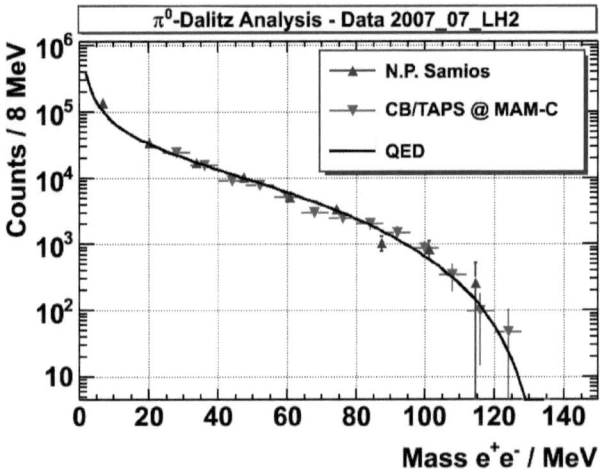

Figure 7.8.: Measured invariant e^+e^--mass distribution for detected π^0-Dalitz events. The red triangles are the data points of this work. The blue triangles are the result of [41] (scaled).

7.5. Conclusion and outlook

In this work the transition form factor of the η-decay into $e^+e^-\gamma$ was measured with higher statistics than ever before.

Beside this, other reactions and decay channels of the mesons π^0, η and ω were successfully investigated.

The only analysis that did not deliver a successful result due to insufficient statistics was the investigation of the ω-Dalitz decay. In order to determine the transition form factor of the ω-meson more data is required; thus more beamtimes with a liquid hydrogen target will have to be performed in the future. It is evident, however, that the statistics would have to be increased by almost two orders of magnitude to provide the statistics comparable to that of the analysis of the η-Dalitz decay in the present work.

7. Results

Appendices

A. Appendix

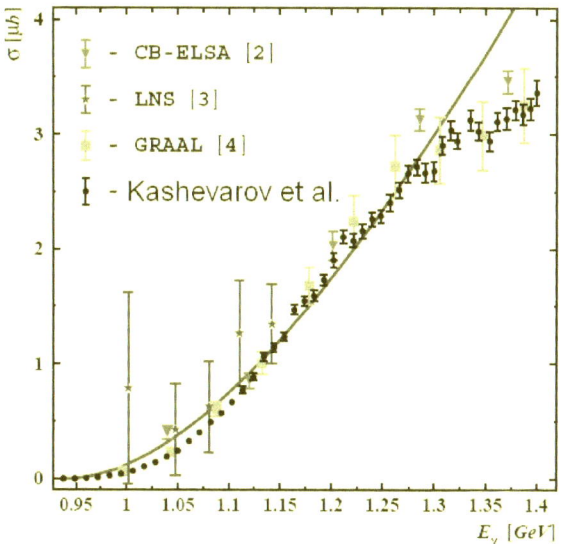

Figure A.1.: Recently published [26]: an investigation of $\eta\pi^0$-production off the proton in the incident energy regime studied in this work.

A. Appendix

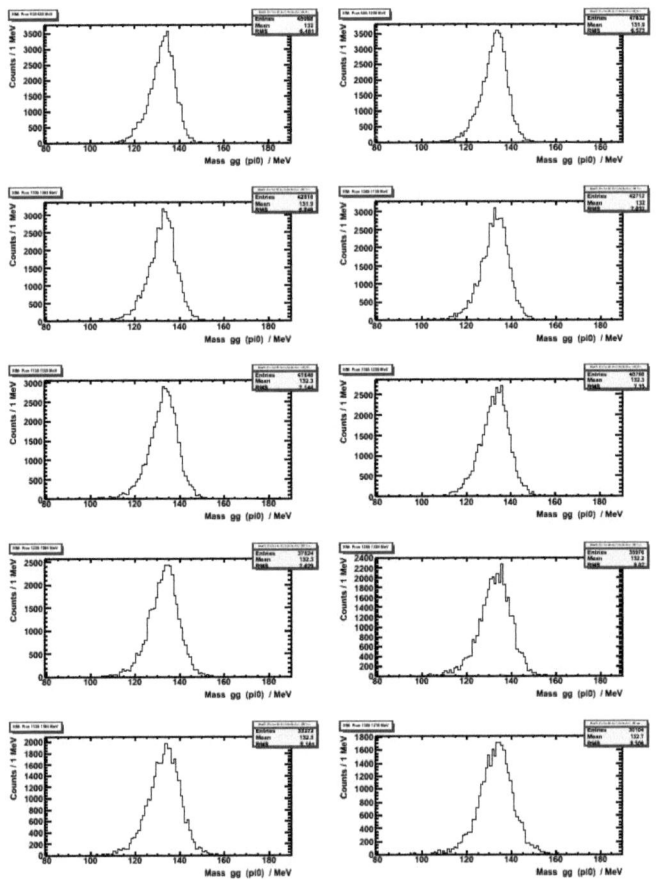

Figure A.2.: Simulation of the $\pi^0\eta$ -Production. The reconstructed invariant π^0-mass is shown for several intervals of incident energy.

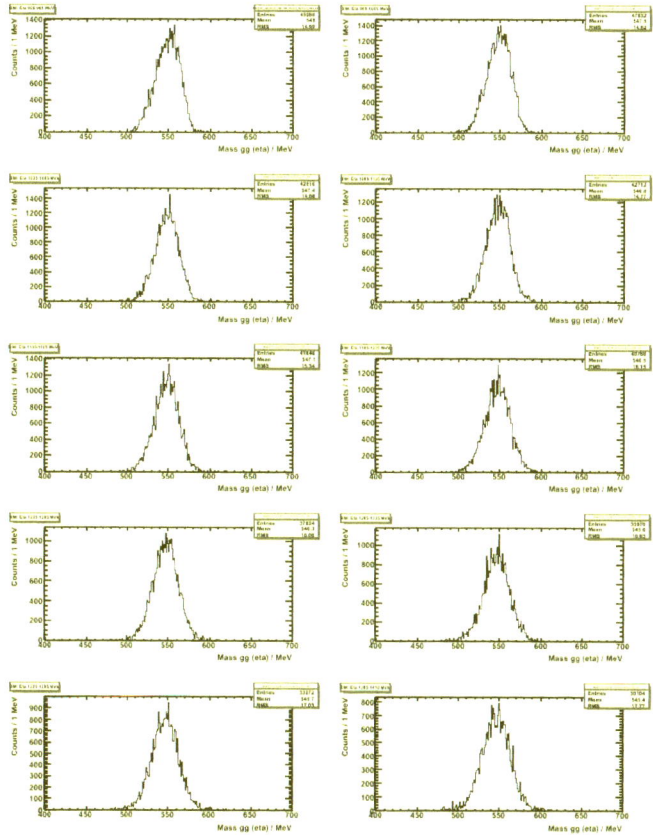

Figure A.3.: Simulation of the $\pi^0\eta$ -Production. The reconstructed invariant η-mass is shown for several intervals of incident energy.

A. Appendix

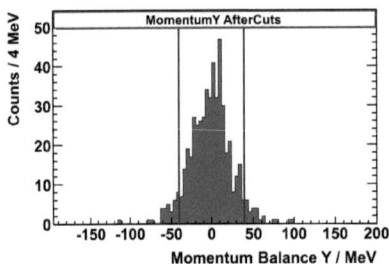

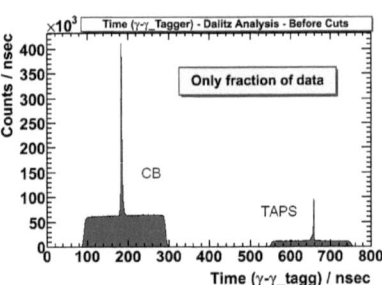

Figure A.4.: Verification of the cut applied on *momentum balance Y* in the η-Dalitz analysis. 1D-Plot of the corresponding variable after cuts.

Figure A.5.: Plot of the timing information of detected photons in the η-Dalitz analysis (before cuts).

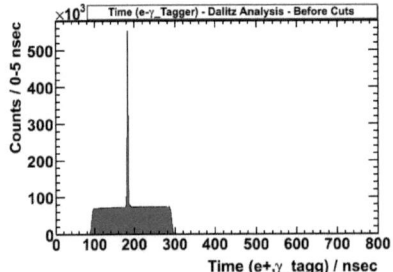

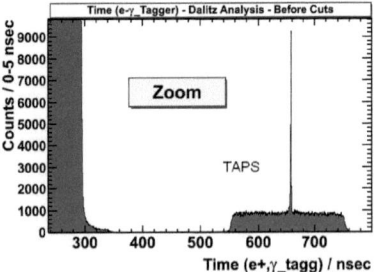

Figure A.6.: Plot of the timing information of first detected charged hit in the η-Dalitz analysis (before cuts).

Figure A.7.: Zoomed plot of the timing information of first detected charged hit in the η-Dalitz analysis (before cuts).

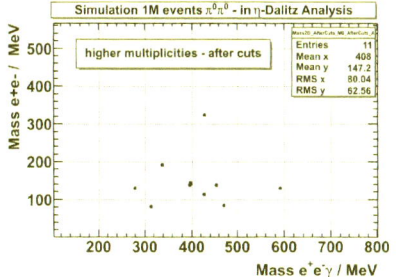
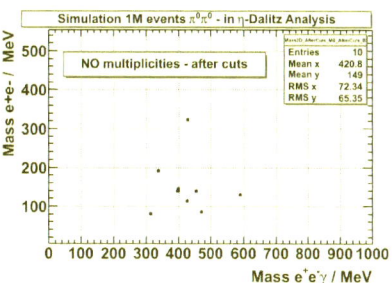

Figure A.8.: Simulation of $\pi^0\pi^0$ in the η-Dalitz analysis. 2D-plot of $m_{e^+e^-}$ against $m_{e^-e^+\gamma}$ of the misidentified events (after cuts). Higher multiplicities have been used.

Figure A.9.: Simulation of $\pi^0\pi^0$ in the η-Dalitz analysis. 2D-plot of $m_{e^-e^-}$ against $m_{e^-e^-\gamma}$ of the misidentified events (after cuts). No higher multiplicities were used.

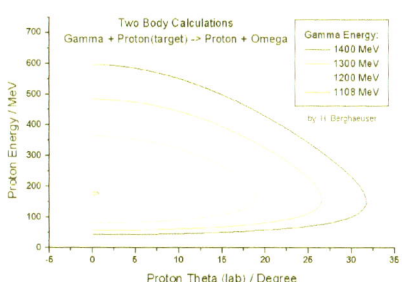
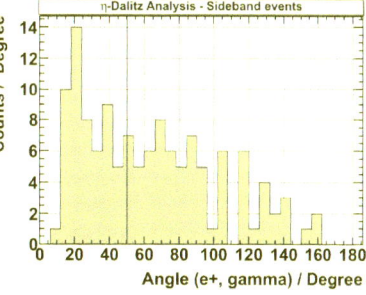

Figure A.10.: Two body calculation for ω-production off the proton. Shown is a plot of the proton energy versus the proton θ-angle.

Figure A.11.: Verification of the cut applied on the $e^{\pm}\gamma$-opening angle in the η-Dalitz analysis. The plotted entries belong to events from the sideband. The 50°-cut removes background. All other cuts were applied (Table 6.3).

A. Appendix

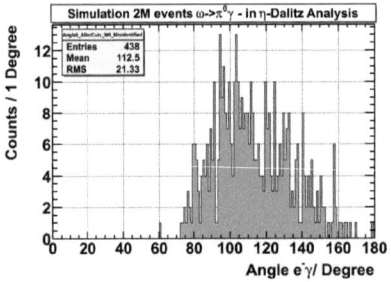

Figure A.12.: Simulation of 2 M events of $\omega \to \pi^0 \gamma$: the reconstructed opening angle of $e^{\pm}\gamma$ of misidentified events (section 6.4.1).

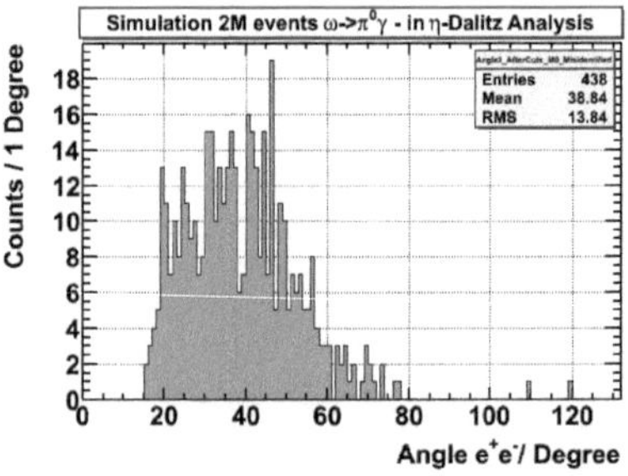

Figure A.13.: Simulation of 2 M events of $\omega \to \pi^0 \gamma$: the reconstructed opening angle of e^+e^- of misidentified events (section 6.4.1).

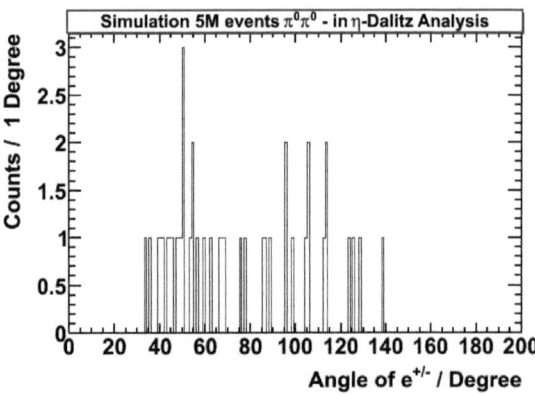

Figure A.14.: Simulation of $\pi^0\pi^0$ in the η-Dalitz analysis. 1D-plot of the θ-angle of electrons surviving the cuts applied in the η-Dalitz analysis.

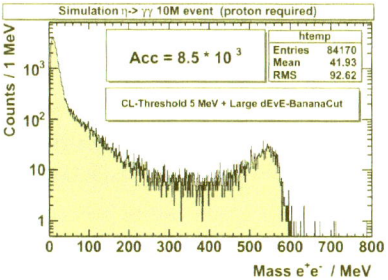
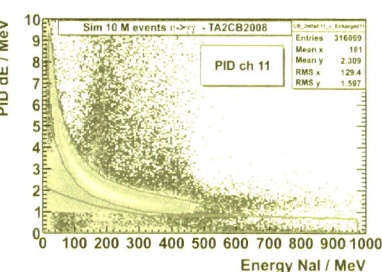

Figure A.15.: Simulation of 10 M events of $\eta \to \gamma\gamma$: misidentified as $e^+e^-\gamma$ (before kinematic cuts, for a lower cluster threshold of 5 MeV and the electron cut shown in Figure A.16).

Figure A.16.: Simulation of 10 M events of $\eta \to \gamma\gamma$: plot of dE versus E (for one PID channel). The proton cut and the *wider* electron cut are shown.

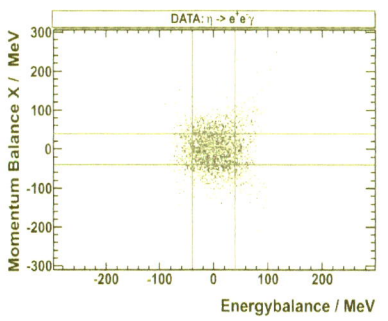
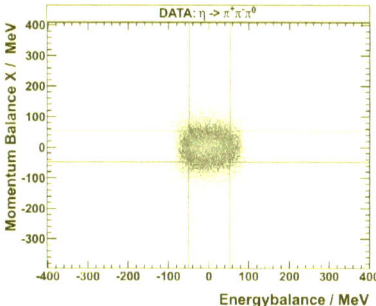

Figure A.17.: Data η-Dalitz analysis: 2D-plot of the momentum balance in X direction versus the energy balance (after all other cuts).

Figure A.18.: Data $\eta \to \pi^+\pi^-\pi^0$-Analysis: 2D-plot of the momentum balance in X direction versus the energy balance (after all other cuts).

A. Appendix

Cut	Cut Range	Counts	η-signal preserved [%]	η-loss [%]
NO CUTS	-	347044	100	0
θ-Proton	max 50°	336374	97	3
Momentum Balance X	-40 <-> 40	250816	72	28
Momentum Balance Y	-40 <-> 40	253291	73	27
Momentum Balance Z	-100 <-> 105	254448	73	27
Energy Balance	-40 <-> 40	203144	59	41
Missing Mass	910 <-> 975	221802	64	36
Beam Energy	750 <-> 1210	246874	71	29
Coplanarity	168 <-> 192	295927	85	15
Opening Angle γ, e^-	50 <-> 175	322229	93	7
Opening Angle γ, e^+	50 <-> 175	298996	86	14
Opening Angle e^+, e^-	19 <-> 140	277096	80	20
e^- Cluster Size	5 <-> 14	264092	76	24
e^+ Cluster Size	3 <-> 12	302223	87	13

Table A.1.: List of applied cuts in the η-Dalitz analysis. Based on a simulation of 2.5 million η-Dalitz events the relative strength of each cut was tested (see section 6.1.3). Since cuts are not independent the overall intensity reduction due to cuts is not the product of all factors. The detection efficiency for the η-Dalitz channel (after cuts) was determined as 1.3 %.

Material	Thickness	Radiation Length X_0 [cm]	$e^{-7x/9X_0}$	Thickness / X_0
lH_2	2 cm	866	0.998205358	$2.3 \cdot 10^{-3}$
Kapton	125 μm	28.6	0.999660119	$4.4 \cdot 10^{-4}$
Mylar	8 μm	28.7	0.99997832	$2.8 \cdot 10^{-5}$
Al	2 μm	8.9	0.999982522	$2.2 \cdot 10^{-5}$
C	1 mm	18.8	0.99587143	$5.3 \cdot 10^{-3}$
All (no PID)			0.9937	$8.1 \cdot 10^{-3}$
PID	2 mm	42.4	0.996337958	$4.7 \cdot 10^{-3}$

Table A.2.: Materials (and their radiation length) in the target region. The 4th column lists the probability for a γ to pass through the corresponding medium (without conversion).

Mass$_{e^+e^-}$ [MeV]	Counts Acc. Corrected	Error Counts Acc. Corrected	$|F_\eta(q_2;0;m_\eta^2)|^2$ [GeV^{-2}]	$\triangle |F_\eta(q_2;0;m_\eta^2)|^2$ [GeV^{-2}]
45	9860	1344	1.04	0.14
75	5323	1109	0.99	0.21
105	3507	425	0.97	0.12
135	3053	311	1.18	0.12
165	1642	276	0.86	0.14
195	1639	202	1.14	0.14
225	1267	195	1.18	0.18
255	1188	170	1.49	0.21
285	896	168	1.57	0.29
315	503	145	1.26	0.36
345	417	108	1.56	0.40
390	226	98	1.67	0.74
450	137	76	3.97	2.20

Table A.3.: Data points obtained in the η-Dalitz analysis. See section 6.24 (Fig. 6.56-6.58) and 7.3 (Fig. 7.4-7.6).

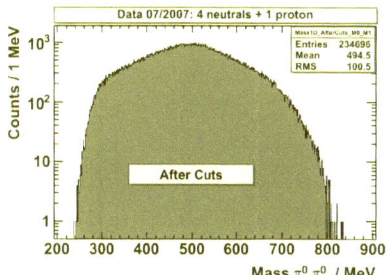

Figure A.19.: Data: analysis of $\pi^0\pi^0$-events after application of cuts listed in Table A.4 and Table 6.21. In total 235000 events were reconstructed (only data from the beamtime in July 2007).

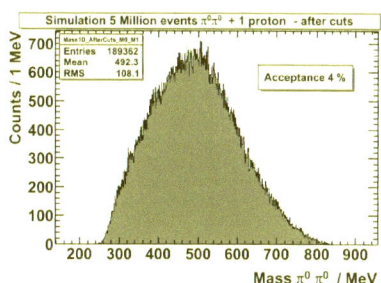

Figure A.20.: Simulation of 5 million $\pi^0\pi^0$-events. The Figure shows the invariant mass of the reconstructed events after cuts (Table A.4). The acceptance is 4%.

A. Appendix

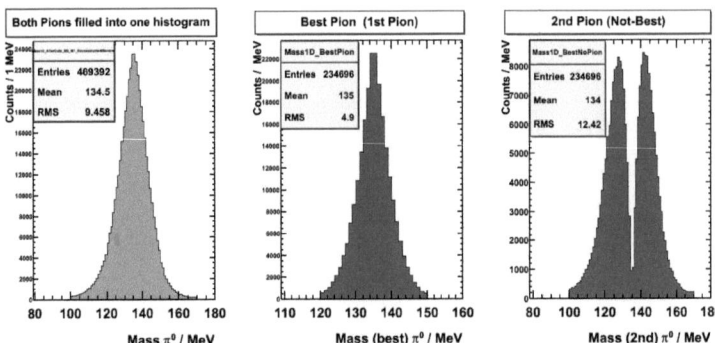

Figure A.21.: Data: analysis of $\pi^0\pi^0$-events; shown are the invariant mass spectra of the best π^0 (middle), the 2nd π^0 (right), and both filled into the same histogram (left) - (after cuts).

Cut	Min	Max
BeamEnergy	750.0	1210.0
Momentum-X	-40.0	40.0
Momentum-Y	-40.0	40.0
Momentum-Z	-100.0	105.0
Missing Mass	910.0	975.0
Coplanarity	168.0	192.0
Best Pion Mass	120	150
Second Pion Mass	100	170

Table A.4.: Applied cuts in the analysis of $\pi^0\pi^0$-events.

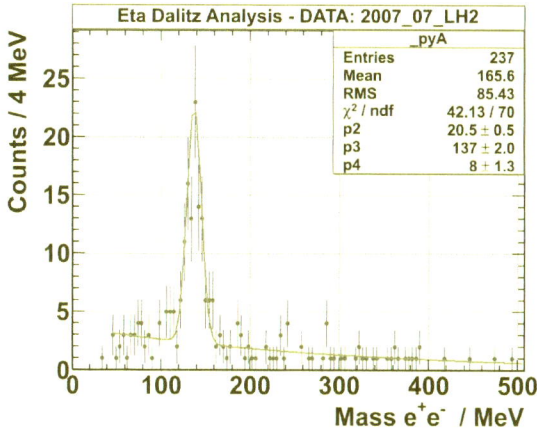

Figure A.22.: Data: misidentified π^0-mesons in the analysis of $\eta \to e^+e^-\gamma$ after cuts (see section 6.2.10).

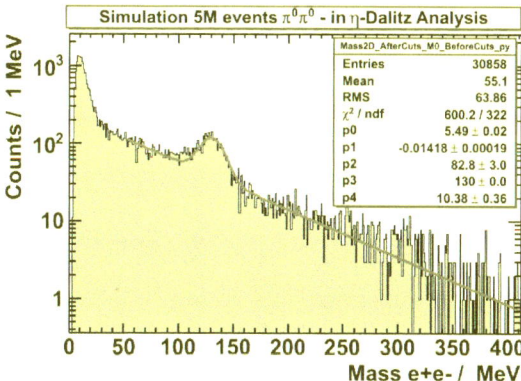

Figure A.23.: Simulation of 5 million $\pi^0\pi^0$-events in the η-Dalitz analysis: misidentified pions with a fit to the π^0-signal (before cuts).

A. Appendix

B List of Figures

1.1. The lightest baryon octet. The figure is taken from [52]. 8

1.2. The lightest baryon decuplet. The figure is taken from [53]. 8

1.3. The meson octet for $J^P = 0^-$. 10

1.4. Illustration of the effective potentials in the case of a) no symmetry breaking and b) spontaneous symmetry breaking. The coordinates x and y correspond to the fields $\sigma, \vec{\pi}$ of the strong interaction. 13

1.5. Cut through the three dimensional potential. In case of an explicit breaking of the symmetry, the potential becomes tilted. 14

1.6. Temperature and density dependence of the chiral condensate corresponding to the Nambu-Jona-Lasinio model [28]. 15

1.7. Feynman graph of an electron scattering process. The form factor is measured in the space-like area ($q^2 < 0$). 16

1.8. Feynman graph of a pair production. The form factor is measured in the time-like area ($q^2 > 0$). 16

1.9. Feynman graph of the η-Dalitz decay. The transition form factor is determined at the $\eta\gamma\gamma^*$-vertex. In this picture the η-Dalitz decay is shown in the picture of the QED. 17

1.10. The η-Dalitz decay in the Vector Meson Dominance model. The virtual γ^* couples to a virtual vector meson. 17

1.11. A scheme plotting the characteristics of the form factor for charged pions. The space-like area of negative q^2 can be measured via electron scattering; whereas the time-like area ($q^2 > 2m_\pi$) can be investigated in annihilation experiments. The shaded area can not be investigated in experiments. This Figure and Figure 1.12 are taken from [31]. 18

B. List of Figures

1.12. Measured form factor of the ω-meson in the time-like regime. For $0 < q < m_\omega - m_{\pi^0}$ the form factor can be measured via the analysis of the ω-Dalitz decay; whereas in the area of $q > m_\omega + m_{\pi^0}$ this investigation can be realized by analyzing the production of $\pi^0 + \omega$ in (e^+e^-)-annihilation experiments [31]. Form factors in the area, that is kinetically forbidden (shaded), can be calculated via a dispersion-relation [4]. 18

1.13. Calculation of the transition form factor of the η-Dalitz decay. 19

1.14. Feynman graph for the investigation of the $\pi^\pm$-form factor in the annihilation of e^+e^-. 19

1.15. Measured time-like π form factor in e^+e^- annihilation (Figure 1.14) compared to a VMD prediction [28],[5]. 19

1.16. Transition form factor of $\eta \to e^+e^-\gamma$ measured by the SND experiment [1]. 21

1.17. Recent result from the heavy ion experiment NA60 [12]. 21

1.18. Measurement of the form factor of $\eta \to \mu^+\mu^-\gamma$ by Lepton-G. The solid line is the fit to the data; the dashed curve presents the VMD prediction [31], [8], [13]. 22

1.19. measurement of the form factor of $\omega \to \mu^+\mu^-\pi^0$ by Lepton-G. The solid line is the fit to the data; the dashed present the VMD prediction [31], [8], [13]. 22

2.1. Floor plan of the MAMI-C accelerator facility including the experimental halls A1, A2, A4, X1. 24

2.2. A RTM showing the increased path radius with increasing energy. 24

2.3. The MAMI-C accelerator. The LINAC, three RTMs and the HDSM are used to accelerate the electron beam up to 1508 MeV. 26

2.4. The Glasgow Tagging system. 27

2.5. Distribution of photon beam; 28

2.6. Pictures taken from [49]. 29

2.7. The hydrogen target cell in a technical drawing. Picture was taken from [49]. 30

2.8. The Crystal Ball. 34

2.9. Segmentation of the Crystal Ball. The major triangles bordered by a thick black line contain the minor triangles. Each minor triangle corresponds to a Na(Tl)I crystal. 35

2.10. The Particle Identification detector. Pictures taken from [49]. 36

2.11.	MWPCs: Wire chamber diagram showing cathode winding and anode wires.	38
2.12.	The TAPS Detector [34].	39
2.13.	A single TAPS crystal together with a veto detector (right), a light guide and a photomultiplier tube.	40
2.14.	Arrangement of the BaF_2 crystals in TAPS seen in beam direction.	40
2.15.	Left: 4 $PbWO_4$ crystals as composition. Right: the new inner Rinf of TAPS.	42
2.16.	Rates in TAPS. The closer a crystal is to the beamline, the higher the rate.	42
2.17.	Photograph of the aluminum frame with all veto detectors and the light guides visible.	42
2.18.	Scheme of the CB readout system.	44
2.19.	Crosstalk measurement; the BaF_2-crystals (x-axis) are plotted against the maximum crosstalk in the TDC-channels (y-axis).	45
2.20.	The TAPS computer system	46
2.21.	Trigger logic of the combined CB/TAPS system.	48
3.1.	ADC spectrum after using a $^{241}Am/^9Be$ radioactive source for energy (gain) calibration [50].	53
3.2.	Reconstructed π^0 mass per channel of the NaI after the first iteration.	55
3.3.	Reconstructed π^0 mass per channel of the NaI after the 27th iteration.	55
3.4.	The fits to the π^0 mass distributions of 50 NaI channels.	56
3.5.	Result of the linear calibration procedure. The η-mass is still off.	57
3.6.	Reconstructed η mass per channel after the first iteration.	59
3.7.	Reconstructed η mass per channel after the 9th iteration.	59
3.8.	Reconstructed η mass after calibration.	59
3.9.	Reconstructed π^0 mass after calibration.	59
3.10.	The *time walk*. For a lower signal the trigger threshold is reached later in time.	60
3.11.	*Time Walk* of NaI-Channel 69.	61
3.12.	Time Walk of all channels plotted versus the NaI-Energy after correction.	61

B. List of Figures

3.13. Time resolution of two neutral hits in the CB. The FWHM of the fit to the peak is 3.04 ns. 62

3.14. Distribution of energy deposit by cosmic Myons on the passage through a BaF_2 crystal. 63

3.15. Invariant mass positions of $m_{\pi^0 meas}$ per channel after the fourth iteration. 64

3.16. Invariant mass positions of $m_{\pi^0 meas}$ per channel after the final 18th iteration. 64

3.17. After the linear calibration, the η-mass is still off. 65

3.18. Deviation of the measured η-mass from the PDG value in the case without (a) and with (b) the second order correction. 67

3.19. After accomplishment of the second order correction: Reconstructed invraiant mass of $\eta \rightarrow \gamma\gamma$ (with TAPS after cuts). 67

3.20. After calibration: time difference of two neutrals detected in the TAPS detector. 69

3.21. PID: a) the wrong correlation in ϕ leads to false detection of a charged particle as neutral hit. b) PID elements and NaI are aligned correctly. Thus the detection of the charged particle works properly. The picture was taken from [32]. 70

3.22. Azimuthal ϕ angle of the NaI hits for coincident hits in the NaI calorimeter and the PID versus the responding PID element ID. 71

3.23. All azimuthal ϕ angles of the NaI calorimeter have been correctly assigned to the corresponding PID elements. 71

3.24. Simulation of $\eta \rightarrow e^+e^-\gamma$: PID energy plotted versus the NaI cluster energy. The 'banana bands' are used to identify protons and $e+/e-$. ... 71

3.25. VETO energy plotted versus the BaF_2 cluster energy. Data from the July run in 2007; after calibration. 72

3.26. Left: Fit of the proton distributions of all 384 VETO elements. Right: A 2D histogram showing the Veto dE versus the BaF_2 energy. The red and the green curve are theoretical Bethe Bloch calculations and are used in the calibration process. 73

3.27. VETO channels plotted versus the BaF_2 channels. Some VETOs are not properly correlated and thus are not on the diagonal line. 74

3.28. The so called HitMatrix of TAPS for charged particles, helps to identify mixed up VETO channels. The shown spectrum does not contain any mixed up channels any more (DATA). 74

3.29. Plot of the calculated TAGGER Photon-Energy against the TAGGER channels. 75

3.30. 2D Spectrum plotting the TAGGER time against the TAGGER channels. 75

3.31. Electron hits in each Tagger channel. The channel number can be converted into a photon energy. Channel 0 corresponds to the highest photon energy. 76

3.32. Tagging efficiency for each TAGGER channel (top) and with background correction (bottom). 79

3.33. Mass$_{e^-e^-\gamma}$ plotted versus Momentum$_{e^-e^-\gamma}$ for simulated $\eta \to e^+e^-\gamma$ events. 81

3.34. Analysis of $\gamma\gamma$ events of experimental data. 81

3.35. Plot of the reconstructed invariant π^0-mass versus the momentum of the π^0. 82

3.36. Plot of the reconstructed invariant η-mass versus the η-momentum. . . . 82

3.37. Invariant η-mass after cuts. Experimental data from TAPS and CB. . . . 82

4.1. The AcquRoot(System) data storage and analysis system. Extracted from [3]. 84

4.2. A schematic representation of the principle structure of the $AR_{HB}2vX$ program line. The output of a $AR_{HB}2vX$ program can be further analyzed or used by calibration macros or the HBAnalysis1v8. 87

4.3. Schematic illustration of the structure of th HBAnalysis1v8 program. . . 90

4.4. Detector geometry implemented in GEANT-4. 93

4.5. The CB/TAPS detector systems in the GEANT geometry of the A2-Sim program. 93

4.6. Server/Computer infrastructure of the A2-Group of the University of Giessen. The blue lines and the black boxes are the 1GBit Ethernet network. See Table 4.1. 97

4.7. Data: 2D-plot showing the energy of PID channel 23 versus the energy of the NaI calorimeter. Furthermore a graphical proton cut (black), π^+/π^--cut (red) as well es an e^+/e^--cut can be seen. 100

4.8. From simulation: a 2D histogram plotting the *time of flight* (TAGGER-Time minus TAPS-Time) against the BaF_2 energy. Also the 2D proton *tof-cut* is shown (black line). 102

B. List of Figures

4.9. From simulation: a 2D histogram plotting the VETO energy against the BaF_2 cluster energy. A proton band as well as an $e+/e-$ can be seen. . . 102

4.10. Simulation of η-Dalitz events. 2D-Plot of the *momentum balance in X* versus the *energy balance*. 103

4.11. Simulation of $\eta \to \pi^+\pi^-\pi^0$ in η-Dalitz analysis. 2D-Plot of the *momentum balance in X* versus the *energy balance* (before cuts). 103

4.12. Simulation of $\eta \to \pi^+\pi^-\pi^0$ in η-Dalitz analysis before cuts. Plot of the *missing mass*. 104

4.13. From simulation: Cluster size of detected electrons/positrons. This picture has been taken from [6]. 104

4.14. From simulation: Cluster size of detected π^+/π^-. This picture has been taken from [6]. 104

4.15. Simulation: $\eta \to \pi^+\pi^-\pi^0$ in the η-Dalitz analysis before cuts. 105

4.16. Simulation: $\eta \to \pi^+\pi^-\pi^0$ in the η-Dalitz analysis after cuts. 105

5.1. Plot of the η-energy against the energy of the proton. 109

5.2. The energy of the backscattered proton is plotted against the proton θ-angle. 109

5.3. Calculation for different incident energies: θ-angle of the proton plotted versus the θ-angle of the η-meson. 109

5.4. Calculation for different incident energies: θ proton plotted versus θ of the ω-meson. 110

5.5. Calculation for different incident energies: θ proton plotted versus energy of the ω-meson. 110

5.6. Calculation for different incident energies: θ-angle of the produced π^0 plotted versus its energy. 110

5.7. Calculation for different incident energies: θ-angle of the backscattered proton plotted versus its energy. 110

5.8. Calculation for different incident energies: the plot shoes the θ-angle of the produced π^0 versus the θ-angle of the backscattered proton. 111

5.9. θ-angle of the produced η' plotted versus its energy. 112

5.10. θ-angle of η' plotted against the θ-angle of the backscattered proton. . . . 112

5.11. Phase space start distribution: Energies of the γ-beam following a $1/E$-distribution. 113

5.12. Phase space start distribution: invariant e^+e^--mass. 113

5.13. Pluto start distribution: γ-beam flux following a $1/E_\gamma$ energy distribution. 114

5.14. Pluto start distribution: the invariant e^+e^--mass of the generated η-Dalitz decays. 114

5.15. Pluto start distribution: the distribution in θ of the generated η-mesons. 114

5.16. Pluto start distribution: the distribution in ϕ of the generated η-mesons. 114

5.17. Pluto start distribution: γ-beam flux following a $1/E_\gamma$ energy distribution. 115

5.18. Pluto start distribution: the invariant e^+e^--mass of the generated ω-Dalitz decays. 115

5.19. Pluto start distribution: γ-beam flux following a $1/E_\gamma$ energy distribution. 115

5.20. Pluto start distribution: the invariant e^+e^--mass of the generated π^0-Dalitz decays. 115

5.21. Phase Space: invariant mass of $e^+e^-\gamma$. 116

5.22. Pluto: invariant mass of $e^+e^-\gamma$. 116

5.23. Phase Space: invariant mass of e^+e^- plotted against $m_{e^+e^-\gamma}$. 116

5.24. Pluto: invariant mass of e^+e^- plotted against $m_{e^+e^-\gamma}$. 116

5.25. Phase Space: invariant mass of $e^+e^-\gamma$. 117

5.26. Pluto: invariant mass of $e^+e^-\gamma$. 117

5.27. Phase Space: reconstructed opening angle of (e^+, e^-). 117

5.28. Pluto: reconstructed opening angle of (e^+, e^-). 117

5.29. Phase Space: reconstructed opening angle of (e^-, γ). 118

5.30. Pluto: reconstructed opening angle of (e^-, γ). 118

5.31. Phase Space: reconstructed opening angle of (e^+, γ). 118

5.32. Pluto: reconstructed opening angle of (e^+, γ). 118

5.33. Phase Space: reconstructed θ-angle of the electron. 119

5.34. Pluto: reconstructed θ-angle of the electron. 119

5.35. Phase Space: reconstructed θ-angle of the positron. 119

5.36. Pluto: reconstructed θ-angle of the positron. 119

5.37. Phase Space: reconstructed θ-angle of the photon. 119

5.38. Pluto: reconstructed θ-angle of the photon. 119

B. List of Figures

5.39. Phase Space: reconstructed θ-angle of the proton. The reduced yield near $\theta = 20°$ is due to the gap between TAPS and CB. 120

5.40. Pluto: reconstructed θ-angle of the proton. 120

5.41. Phase Space: reconstructed ϕ-angle of the electron. The reduced yield near $\phi = 0°$ is due to the CB support structure. 120

5.42. Pluto: reconstructed ϕ-angle of the electron. 120

5.43. Phase Space: reconstructed ϕ-angle of the positron. 121

5.44. Pluto: reconstructed ϕ-angle of the positron. 121

5.45. Phase Space: reconstructed ϕ-angle of the photon. 121

5.46. Pluto: reconstructed ϕ-angle of the photon. 121

5.47. Phase Space: reconstructed ϕ-angle of the proton. 121

5.48. Pluto: reconstructed ϕ-angle of the proton. 121

5.49. Phase Space: 2D-Plot of θ-proton versus the θ-angle of the produced η-meson. 122

5.50. Pluto: 2D-Plot of θ-proton versus the θ-angle of the produced η-meson. . 122

5.51. The energy information of the CB and TAPS detectors in the Monte Carlo simulation with and without modified correction factor. 122

5.52. MC-simulation: invariant $e^+e^-\gamma$-mass distribution after a proper scaling of the energy, the η-mass is located at 547.8 MeV (fit). 122

6.1. MC-Simulation of 10.000.000 events of $\eta \to \gamma\gamma$. 127

6.2. Simulation of $\eta \to \gamma\gamma$ (detection of proton required). The cluster thresholds of CB and TAPS was set to 20 MeV. 128

6.3. Simulation of $\eta \to \gamma\gamma$ (detection of proton required). The cluster thresholds of CB and TAPS was set to 50 MeV. 128

6.4. Simulation of $\eta \to \gamma\gamma$ (no proton required). The cluster thresholds of CB and TAPS was set to 20 MeV. 128

6.5. Simulation of $\eta \to \gamma\gamma$ (no proton required). The cluster thresholds of CB and TAPS was set to 50 MeV. 128

6.6. Simulation: reconstructed invariant mass spectrum of $3\pi^0$ after cuts. . . . 129

6.7. Distribution of the detected e^+e^--mass in the η-Dalitz analysis of simulated data. 130

6.8. Determined acceptance of $\eta \to e^+e^-\gamma$. 130

6.9. Fit to the acceptance histogram. 131

6.10. Histogram showing both, the corrected and the original acceptance, respectively. 131

6.11. The corrected acceptance used in the η-Dalitz analysis. 132

6.12. Corrected acceptance with a binning of 4 MeV. 132

6.13. Investigation of the shape of the acceptance for different cluster threshold and different cuts on the opening angle of e^+e^-. 132

6.14. Investigation of the split-off-effect. Five million events of $\eta \to e^+e^-\gamma$ (proton) were simulated and thereafter analyzed using a function only investigating the final state $e^+e^-\gamma\gamma$. 134

6.15. Simulation of 0.9M events of $\eta \to \pi^+\pi^-\gamma$: spectrum of the reconstructed invariant mass. 135

6.16. Simulation: background in the analysis of $\pi^+\pi^-\gamma$ stemming from $\pi^+\pi^-\pi^0$. 135

6.17. Simulation of $\eta \to \pi^+\pi^-\gamma$. In total 18% enter in the η-Dalitz analysis (before cuts). 135

6.18. Simulation: Invariant mass spectrum of $\eta \to \pi^+\pi^-\pi^0$ (proton) after cuts. 135

6.19. Simulation: reconstructed invariant $\pi^0\gamma$-mass (no cuts). 137

6.20. Simulation: reconstructed invariant $\pi^0\gamma$-mass after a strict cut on the π^0-mass. 137

6.21. Simulation: reconstructed invariant $\pi^0\gamma$-mass after an additional cut on the θ-angle of the detected proton. 138

6.22. Simulation: reconstructed invariant $\pi^0\gamma$-mass after an additional cut on the coplanarity. 138

6.23. Simulation: reconstructed invariant $\pi^0\gamma$-mass after application of cut-setting B. 139

6.24. Simulation: reconstructed invariant $\pi^0\gamma$-mass after application of cut-setting C. 139

6.25. Simulation: reconstructed invariant $\pi^0\gamma$-mass after application of cut-setting A. 140

6.26. Simulation of $\omega \to \pi^0\pi^+\pi^-$: reconstructed ω-mass after cuts. 140

6.27. Sim: invariant mass of reconstructed events in the analysis of $\omega \to e^+e^-\pi^0$. 142

6.28. Sim: 2D plot of mass$_{e^+e^-}$ versus the mass$_{e^+e^-\pi^0}$ in the analysis of the ω-Dalitz decay. 142

B. List of Figures

6.29. Sim: distribution of the invariant e^+e^--mass of reconstructed ω-events. . 142

6.30. Sim: acceptance of ω-Dalitz detection depending on $m_{e^+e^-}$. 142

6.31. Simulation of π^0-Dalitz using a cluster threshold of 20 MeV before cuts. . 144

6.32. Simulation of π^0-Dalitz using a cluster threshold of 20 MeV after cuts; the red ellipse marks the problematic regime. 144

6.33. Simulation of π^0-Dalitz; invariant mass$_{e^+e^-\gamma}$ of reconstructed reconstructed events (after cuts and threshold of 50 MeV). 144

6.34. Simulation of π^0-Dalitz using a cluster threshold of 50 MeV after cuts. . . 144

6.35. Simulation of π^0-Dalitz; invariant mass$_{e^+e^-}$ of reconstructed events (after cuts). 145

6.36. Acceptance of the π^0-Dalitz decay; the red line shows the corrected shape of the acceptance (see text). 145

6.37. Plot of the timing information of detected protons in the η-Dalitz analysis (before cuts). 146

6.38. Plot of the timing information of detected proton in the η-Dalitz analysis (after cuts, except for the cut on the proton timing). 146

6.39. Plot of the time of flight information of detected protons (TAPS) in the η-Dalitz analysis (before cuts). 147

6.40. Plot of the time of flight information of detected protons (TAPS) in the η-Dalitz analysis (after cuts). The used TOF-banana cut is illustrated by the red curve. 147

6.41. DATA: reconstructed invariant $\gamma\gamma$-mass after cuts in the exclusive η-analysis. 148

6.42. DATA: reconstructed invariant $3\pi^0$-mass after cuts. 148

6.43. DATA: reconstructed invariant $\pi^0\gamma\gamma$-mass after cuts in the exclusive analysis of $\eta \to \pi^0\gamma\gamma$. Red curve is the fit to the data (combination of two Gaussian functions). 150

6.44. DATA: reconstructed invariant $\pi^0\gamma\gamma$-mass after cuts in the exclusive η-analysis. The dashed curve is the fit to the background (Gaussian). . . . 150

6.45. DATA: reconstructed invariant $e^+e^-\gamma$-mass before cuts in the exclusive analysis of the η-Dalitz decay 152

6.46. DATA - η-Dalitz analysis: Projection of $m_{e^+e^-}$ onto $m_{e^+e^-\gamma}$ for the interval of 120 MeV to 150 MeV. 152

6.47. DATA: reconstructed invariant $e^+e^-\gamma$-mass after cuts in the exclusive analysis of the η-Dalitz decay (Beamtime June 2007). 152

6.48. DATA: reconstructed invariant $e^+e^-\gamma$-mass after cuts in the exclusive analysis of the η-Dalitz decay (Beamtime July 2007). 152

6.49. DATA - η-Dalitz analysis: 2D-plot of the reconstructed invariant e^+e^--mass versus $m_{e^+e^-\gamma}$ after cuts (Beamtime June 2007). 153

6.50. DATA - η-Dalitz analysis: 2D-plot of the reconstructed invariant e^+e^--mass versus $m_{e^+e^-\gamma}$ after cuts (Beamtime July 2007). 153

6.51. DATA: reconstructed invariant $e^+e^-\gamma$-mass after cuts in the exclusive analysis of the η-Dalitz decay (both beamtimes added). 153

6.52. DATA - η-Dalitz analysis: 2D-plot of the reconstructed invariant e^+e^--mass versus $m_{e^+e^-\gamma}$ after cuts (both beamtimes added). 153

6.53. η-Dalitz analysis: Projections of the 2D-plot of Figure 6.52 onto the $m_{e^+e^-\gamma}$-axis for slices in $m_{e^+e^-}$ with a width of 30 MeV. 155

6.54. η-Dalitz analysis: Projections of the 2D-plot of Figure 6.52 onto the $m_{e^+e^-\gamma}$-axis for slices in $m_{e^+e^-}$ with a width of 30 MeV. 156

6.55. DATA: reconstructed invariant e^+e^--mass after cuts and background subtraction without acceptance correction (both beamtimes 2007). See Table 6.17. 157

6.56. DATA - η-Dalitz analysis: reconstructed invariant mass of e^+e^- after acceptance correction. The dashed black line is the scaled QED-curve. The black solid line is a fit to the data points within the VMD-model. . . . 157

6.57. DATA - η-Dalitz analysis: Transition form factor of the η-meson. The red curve is the fit to the data. 158

6.58. DATA - η-Dalitz analysis: Transition form factor of the η-meson in comparison to theoretical predictions (zoomed view). 158

6.59. Verification of the missing mass cut applied in the η-Dalitz analysis. Plot of the missing mass before cuts. 158

6.60. Verification of the missing mass cut applied in the η-Dalitz analysis. Plot of the missing mass after all other cuts. 158

6.61. Verification of the coplanarity cut applied in the η-Dalitz analysis. Plot of the coplanarity after all other cuts. 159

6.62. Verification of the cut applied on the θ-proton in the η-Dalitz analysis. Plot of the θ-angle of the proton after all other cuts. 159

B. List of Figures

6.63. Verification of the cut applied on *momentum balance X* in the η-Dalitz analysis. 1D-Plot of the corresponding variable after all other cuts. . . . 160

6.64. Verification of the cut applied on *momentum balance Z* in the η-Dalitz analysis. Plot of the corresponding variable after all other cuts. 160

6.65. Verification of the cut applied on *cluster size* in the η-Dalitz analysis. 1D-Plot of the corresponding variable after all other cuts. 160

6.66. Verification of the cut applied on *cluster size* in the η-Dalitz analysis. Plot of the corresponding variable after all other cuts. 160

6.67. Verification of the cut applied on the e^+e^--opening angle in the η-Dalitz analysis. 1D-Plot of the corresponding variable before cuts. 161

6.68. Verification of the cut applied on the e^+e^--opening angle in the η-Dalitz analysis. Plot of the corresponding variable after all other cuts. 161

6.69. Verification of the cut applied on the $e^{+/-}\gamma$-opening angle in the η-Dalitz analysis. 1D-Plot of the corresponding variable before all cuts cuts. . . . 161

6.70. Verification of the cut applied on the $e^{+/-}\gamma$-opening angle in the η-Dalitz analysis. Plot of the corresponding variable after all cuts cuts. 161

6.71. Invariant mass of all detected $\pi^+\pi^-\pi^0$-events before cuts. 163

6.72. Invariant mass of $\pi^+\pi^-\pi^0$-events after the application of cuts on the kinematics. 163

6.73. Invariant mass of $\pi^+\pi^-\pi^0$-events after application of cuts listed in Table 6.4. The number of counts in the η-peak was determined by the fit to the data points. 163

6.74. Comparison of the invariant $\pi^0\gamma$ mass for different cuts (without time cuts). 164

6.75. Comparison of the invariant $\pi^0\gamma$ mass for different cuts (without time cuts). 164

6.76. Invariant $\gamma\gamma\gamma$-mass; only cuts on the timing were applied. 165

6.77. Invariant $\pi^0\gamma$-mass; an additional cut on the π^0-mass was applied. 165

6.78. Invariant $\pi^0\gamma$-mass after a strict cut on the π^0-mass (and cuts on the timing). 166

6.79. Invariant $\pi^0\gamma$-mass after application of the cuts-list 'C'. 166

6.80. Invariant $\pi^0\gamma$-mass after application of the cuts-list 'A'. 166

6.81. Invariant $\pi^0\gamma$-mass after application of the cuts-list 'B'. The number of ω-mesons was determined by a fit to the data points. 167

6.82. Invariant $\pi^0\pi^+\pi^-$-mass after the application of cuts. The background and the ω-signal were fitted. 168

6.83. Invariant mass of $e^+e^-\pi^0$ after all cuts. 169

6.84. 2D-plane of $m_{e^+e^-}$ plotted versus $m_{e^+e^-\pi^0}$ after cuts for the analysis of the ω-Dalitz decay. 169

6.85. Reconstructed π^0 and η-signals with a fit to the η-signal for the interval 935-1035 MeV of the incident energy. 170

6.86. Reconstructed π^0 and η-signals with a fit to the η-signal for the interval 1035-1085 MeV of the incident energy. 171

6.87. Reconstructed π^0 and η-signals with a fit to the η-signal for the interval 1085-1135 MeV of the incident energy. 171

6.88. Reconstructed π^0 and η-signals for the interval 1335-1385 MeV of the incident energy (without a plot of the fit). 172

6.89. Shown is the measured cross section for $\pi^0\eta$-production (result of this work). The data points are listed in Table 6.22. 173

6.90. Invariant $\pi^0\gamma$-mass stemming from $\pi^0\pi^0$ and $\pi^0\eta$-events. 175

6.91. π^0-Dalitz analysis: invariant $e^+e^-\gamma$-mass of the reconstructed events after cuts with a fit to the π^0-signal. 176

6.92. Dalitz analysis: 2D-plot of $m_{e^-e^-}$ against $m_{e^+e^-\gamma}$ after cuts. 176

6.93. π^0-Dalitz analysis: projection onto the $m_{e^-e^-}$-axis for the signal (black), the side bands (red, green), and the corrected signal (blue). 176

6.94. π^0-Dalitz analysis: distribution of the invariant e^+e^--mass after background subtraction and after acceptance correction. 176

6.95. Simulation of 10M events of $\eta \rightarrow \gamma\gamma$: plotted are the invariant masses of 2γ-pairs that were misidentified as e^+e^-. 181

6.96. Simulation of 10M events of $\eta \rightarrow \gamma\gamma$: plotted is the opening angle of the detected e^+, e^- before cuts. 181

6.97. Acceptance for the conversion process (black line) as a function of the cut on the e^+e^--opening angle. This acceptance was normalized to the number of produced photons per η-meson (see text). 182

6.98. Contribution in % of the conversion processes to the η-Dalitz channel as a function of the cut on the e^+e^--opening angle (see text). 182

6.99. Simulation of 10M events $\eta \rightarrow \gamma\gamma$: misidentified as $e^+e^-\gamma$ (before kinematic cuts). 182

235

B. List of Figures

6.100 Simulation of 10M events $\eta \to \gamma\gamma$: misidentified as $e^+e^-\gamma$. Only 3 counts survive the whole series of cuts on the kinematics. 182

6.101 Simulation of 10M events $\eta \to \gamma\gamma$: after a conversion of one γ into e^+e^-, the opening angle of the charged lepton pair is rather small. 184

6.102 Data: analysis of η-Dalitz events. A minimum opening angle of 19° is required. 184

6.103 Simulation of 3.6M events $\eta \to \pi^+\pi^-\gamma$: before the cuts are applied in the η-Dalitz analysis. 185

6.104 Simulation of 3.6M events $\eta \to \pi^+\pi^-\gamma$: after the cuts are applied in the η-Dalitz analysis the 2D-plane is empty. 185

6.105 Simulation of 1M events of $\eta \to \pi^0\pi^0\pi^0$: plot of $m_{e^+e^-}$ against $m_{e^+e^-\gamma}$ before cuts (see text). 186

6.106 Simulation of 1M events of $\eta \to \pi^0\pi^0\pi^0$: plot of $m_{e^+e^-}$ against $m_{e^+e^-\gamma}$. No events survived the cuts. 186

6.107 Simulation of 3 M events of $\eta \to \pi^0\gamma\gamma$: plot of $m_{e^+e^-}$ against $m_{e^+e^-\gamma}$ before cuts (see text). 186

6.108 Simulation of 3 M events of $\eta \to \pi^0\gamma\gamma$: plot of $m_{e^+e^-}$ against $m_{e^+e^-\gamma}$. Some events survived the cuts. 186

6.109 Simulation of 5 M events of $\omega \to e^+e^-\pi^0$: plot of $m_{e^+e^-}$ against $m_{e^+e^-\gamma}$ (before cuts). 187

6.110 Simulation of 5 M events of $\omega \to e^+e^-\pi^0$: plot of $m_{e^+e^-}$ against $m_{e^+e^-\gamma}$. Many events survived the cuts. 187

6.111 Simulation of 1 M events of $\omega \to \pi^+\pi^-\pi^0$: plot of $m_{e^+e^-}$ against $m_{e^+e^-\gamma}$ (before cuts). 188

6.112 Simulation of 1 M events of $\omega \to \pi^-\pi^-\pi^0$: plot of $m_{e^+e^-}$ against $m_{e^+e^-\gamma}$. No events survive the cuts. 188

6.113 Simulation of 2 M events of $\omega \to \pi^0\gamma$: plot of $m_{e^+e^-}$ against $m_{e^+e^-\gamma}$ (before cuts). 188

6.114 Simulation of 2 M events of $\omega \to \pi^0\gamma$: plot of $m_{e^+e^-}$ against $m_{e^+e^-\gamma}$. ... 188

6.115 Simulation of 2 M events of $\omega \to \pi^0\gamma$: plot of $m_{e^+e^-}$ against $m_{e^+e^-\gamma}$ (before cuts). The misidentified $\pi^0 \to \gamma\gamma$-events are marked by the red curve of the 2D-cut. 190

6.116 Simulation of 2 M events of $\omega \to \pi^0\gamma$: reconstructed invariant e^+e^--mass of the misidentified events (see text). 190

6.117 Simulation of 0.3M $\pi^0\eta$-events: reconstructed invariant $e^+e^-\gamma$-mass of $\pi^0\eta$-production events. 190

6.118 Simulation of 0.3M $\pi^0\eta$-events: plot of the missing mass. Using an appropriate cut (Table 6.3) leads to a strong reduction of this background channel. 190

6.119 Simulation of 0.3M $\pi^0\eta$-events: invariant mass of e^+e^- plotted versus $m_{e^+e^-\gamma}$ (before cuts). 191

6.120 Simulation of 1.5M $\pi^0\eta$-events: invariant mass of e^+e^- plotted versus $m_{e^-e^-\gamma}$ (after cuts). 191

6.121 Simulation of $\pi^0\pi^0$ in the η-Dalitz analysis. 2D-plot of $m_{e^-e^-}$ against $m_{e^-e^-\gamma}$ of the misidentified events (before cuts). 192

6.122 Simulation of $\pi^0\pi^0$ in the η-Dalitz analysis. 2D-plot of $m_{e^+e^-}$ against $m_{e^-e^+\gamma}$ of the misidentified events (before cuts). 192

6.123 Simulation of $\pi^0\pi^0$ in the η-Dalitz analysis. 1D-plot of the θ-angle of electrons in the η-Dalitz final state (before cuts). 193

6.124 Simulation of $\pi^0\pi^0$ in the η-Dalitz analysis. 2D-plot of $m_{e^-e^-}$ against $m_{e^-e^-\gamma}$ of the misidentified events (after cuts plus extra cut on the θ-angle of e^+e^-). 193

6.125 Simulation of $\pi^0\pi^0$ in the η-Dalitz analysis. 2D-plot of $m_{e^-e^-}$ against $m_{e^-e^-\gamma}$ of the misidentified events (after cuts; with a maximum γ-beam energy of 1 GeV). 194

6.126 Simulation of $\pi^0\pi^0$ in the η-Dalitz analysis. 2D-plot of $m_{e^-e^-}$ against $m_{e^-e^-\gamma}$ of the misidentified events (after cuts; with a maximum γ-beam energy of 1 GeV; all cuts on the momentum balance were 25% less strict). 194

6.127 Simulation of $\pi^-\pi^+$ in the η-Dalitz analysis. 194

6.128 Simulation of $\pi^+\pi^-$ in the η-Dalitz analysis. 2D-plot of $m_{e^-e^-}$ against $m_{e^-e^+\gamma}$ (after cuts). No events survive. 194

6.129 Simulation: background in the analysis of $\omega \to \pi^0\gamma$. The production of π^0 off the neutron enters in the same final state, when the neutron is misidentified as γ. 195

6.130 Simulation: π^0-production events (off the neutron) fulfill the mass cut on the best π^0, which is applied in the ω-analysis. 195

6.131 Simulation of $\pi^0 + n$: plot of the missing mass in a 3γ-analysis. 196

B. List of Figures

7.1. Measured total cross section of $\pi^0\eta$-production in photon induced reactions off the proton (for incident energies up to 1408 MeV) in comparison to the result obtained in [26]. 198

7.2. Data: analysis of $\eta \to e^+e^-\gamma$ events. 2D-Plot of the *momentum balance in X* versus the *energy balance*. 199

7.3. Simulation of 3.6 million events $\eta \to \pi^+\pi^-\gamma$ in the η-Dalitz analysis. 2D-Plot of the *momentum balance in X* versus the *energy balance* (before cuts). 199

7.4. Analysis of the η-Dalitz decay; distribution of $m_{e^+e^-}$ of the reconstructed events after acceptance correction. The dotted curve is the QED prediction scaled to the data points below 120 MeV. The solid curve is a fit within the VMD-model using a monopole form factor. 201

7.5. Measurement of the η-Dalitz transition form factor. The red triangles are the data points of this work (the red line is the fit to the data). The black squares show the result of the SND experiment [1]. The green line shows a calculation performed by Terschluesen and Leupold. The black curve is the fit curve to the data of [31]. 204

7.6. Measurement of the η-Dalitz transition form factor. The red triangles are the data points (the red line is the fit to the data). The black points are the result from [12]. The green line shows a calculation performed by Terschluesen and Leupold. The black curve is the fit curve to the data of [31]. 205

7.7. The slope b_η of the η transition form factor. The result of this work is shown in comparison to former results (SND, NA60, Lepton-G). 206

7.8. Measured invariant e^+e^--mass distribution for detected π^0-Dalitz events. The red triangles are the data points of this work. The blue triangles are the result of [41] (scaled). 206

A.1. Recently published [26]: an investigation of $\eta\pi^0$-production off the proton in the incident energy regime studied in this work. 211

A.2. Simulation of the $\pi^0\eta$ -Production. The reconstructed invariant π^0-mass is shown for several intervals of incident energy. 212

A.3. Simulation of the $\pi^0\eta$ -Production. The reconstructed invariant η-mass is shown for several intervals of incident energy. 213

A.4. Verification of the cut applied on *momentum balance Y* in the η-Dalitz analysis. 1D-Plot of the corresponding variable after cuts. 214

A.5. Plot of the timing information of detected photons in the η-Dalitz analysis (before cuts). 214

A.6. Plot of the timing information of first detected charged hit in the η-Dalitz analysis (before cuts). 214

A.7. Zoomed plot of the timing information of first detected charged hit in the η-Dalitz analysis (before cuts). 214

A.8. Simulation of $\pi^0\pi^0$ in the η-Dalitz analysis. 2D-plot of $m_{e^-e^-}$ against $m_{e^-e^-\gamma}$ of the misidentified events (after cuts). Higher multiplicities have been used. 215

A.9. Simulation of $\pi^0\pi^0$ in the η-Dalitz analysis. 2D-plot of $m_{e^-e^-}$ against $m_{e^-e^+\gamma}$ of the misidentified events (after cuts). No higher multiplicities were used. 215

A.10. Two body calculation for ω-production off the proton. Shown is a plot of the proton energy versus the proton θ-angle. 215

A.11. Verification of the cut applied on the $e^\pm\gamma$-opening angle in the η-Dalitz analysis. The plotted entries belong to events from the sideband. The $50°$-cut removes background. All other cuts were applied (Table6.3). . . . 215

A.12. Simulation of 2 M events of $\omega \to \pi^0\gamma$: the reconstructed opening angle of $e^\pm\gamma$ of misidentified events (section 6.4.1). 216

A.13. Simulation of 2 M events of $\omega \to \pi^0\gamma$: the reconstructed opening angle of e^+e^- of misidentified events (section 6.4.1). 216

A.14. Simulation of $\pi^0\pi^0$ in the η-Dalitz analysis. 1D-plot of the θ-angle of electrons surviving the cuts applied in the η-Dalitz analysis. 216

A.15. Simulation of 10 M events of $\eta \to \gamma\gamma$: misidentified as $e^+e^-\gamma$ (before kinematic cuts, for a lower cluster threshold of 5 MeV and the electron cut shown in Figure A.16). 217

A.16. Simulation of 10 M events of $\eta \to \gamma\gamma$: plot of dE versus E (for one PID channel). The proton cut and the *wider* electron cut are shown. 217

A.17. Data η-Dalitz analysis: 2D-plot of the momentum balance in X direction versus the energy balance (after all other cuts). 217

A.18. Data $\eta \to \pi^+\pi^-\pi^0$-Analysis: 2D-plot of the momentum balance in X direction versus the energy balance (after all other cuts). 217

A.19. Data: analysis of $\pi^0\pi^0$-events after application of cuts listed in Table A.4 and Table 6.21. In total 235000 events were reconstructed (only data from the beamtime in July 2007). 219

B. List of Figures

A.20. Simulation of 5 million $\pi^0\pi^0$-events. The Figure shows the invariant mass of the recontructed events after cuts (Table A.4). The acceptance is 4%. ... 219

A.21. Data: analysis of $\pi^0\pi^0$-events; shown are the invariant mass spectra of the best π^0 (middle), the 2nd π^0 (right), and both filled into the same histogram (left) - (after cuts). ... 220

A.22. Data: misidentified π^0-mesons in the analysis of $\eta \to e^+e^-\gamma$ after cuts (see section 6.2.10). ... 221

A.23. Simulation of 5 million $\pi^0\pi^0$-events in the η-Dalitz analysis: misidentified pions with a fit to the π^0-signal (before cuts). ... 221

C List of Tables

1.1.	The elementary particles contained in the standard model. Leptons and Quarks are Fermions, meaning that their Spin is equal to 1/2. Not listed here but also included in the Standard Model are the corresponding antiparticles [44].	4
1.2.	Elementary forces of the Standard Model and their field bosons [44].	4
1.3.	Properties of elementary forces.	6
1.4.	Some properties of interest of the π^0-meson [19].	10
1.5.	Some properties of the η and the ω-meson [19].	11
1.6.	Experiments and results of the analysis of $\eta \to e^+e^-\gamma$ and $\eta \to \mu^+\mu^-\gamma$.	21
2.1.	Main parameters of MAMI as taken from [11].	25
2.2.	Properties of the targets materials.	33
2.3.	The properties of the NaI calorimeter.	34
2.4.	Parameters of the Na(Tl)I crystals.	35
2.5.	Properties of MWPCs.	38
2.6.	Main parameters of the TAPS detector.	41
2.7.	Main parameters of the BaF_2 crystals.	41
2.8.	Main parameters of $PbWO_4$ in comparison to BaF_2.	43
2.9.	Overview on (analyzed) beamtimes.	49
3.1.	Overview over all calibrations performed in Giessen (since 2007); these are available for download [7]. The calibrations marked with a green $\sqrt{}$ were done by the author himself.	52
3.2.	Difference of the reconstructed η-mass to the PDG value (for LH_2-Beamtimes 06/2007 and 07/2007)	55

C. List of Tables

3.3. Photon flux determination of beamtime 07/2007 for certain energy ranges (without dead-time-correction). The broken channels were corrected - values of neighbours were used. 80

3.4. Photon flux without dead time correction for the η-Dalitz analysis (beamtime 07/2007). 81

4.1. List of servers and workstations used for calibrations and analyses. 97

4.2. LOGIC for particle identification in the CB apparatus. 100

4.3. LOGIC of the TAPS (TA2Taps) particle identification. '-' stands for 'not active', '1' means fulfilled, '0' stands for 'not fulfilled'. 101

5.1. Performed MC-Simulations. (*)In this case ten times 0.3 M events were simulated for (ten) different intervals of incident energy. 108

6.1. Applied cuts. 127

6.2. Applied cuts. 129

6.3. Applied cuts in the analysis of simulated and real-data events of $\eta \rightarrow e^+e^-\gamma$ (proton). 133

6.4. Applied cuts in the analysis of simulated events $\eta \rightarrow \pi^+\pi^-\gamma$ (proton). . . 133

6.5. Applied cuts in the exclusive analysis of $\eta \rightarrow \pi^+\pi^-\pi^0$ (simulation). . . . 136

6.6. Very strict 'Cut-Setting A'. 138

6.7. 'Cut-Setting B'. 138

6.8. Very strict 'Cut-Setting C'. 139

6.9. Applied cuts in the exclusive analysis of $\omega \rightarrow \pi^+\pi^-\pi^0$ (simulation). . . . 139

6.10. Applied cuts in the exclusive analysis of $\omega \rightarrow e^+e^-\pi^0$ (simulation). 141

6.11. Applied cuts in the analysis of $\pi^0\eta$. (*)The cut on the beam energy corresponds to each of the intervals (see Table 6.12). 143

6.12. Determined Acceptance of $\pi^0\eta$ for different intervals of incident energy. . 143

6.13. Applied cuts in the exclusive analysis of $\pi^0 \rightarrow e^+e^-\gamma$ (simulation). 145

6.14. Applied *time* cuts in the exclusive analysis of $\eta \rightarrow \gamma\gamma$. 147

6.15. Applied time cuts in the analysis of the data 2007-07$-lH_2$. Positrons refers to the 2nd charged hit, that is not a proton; electron to the first. . 151

6.16. Applied time cuts in the analysis of the first fraction of data of the beamtime 2007-06–lH_2. 151

6.17. η-Dalitz Analysis: Data Points corresponding to Figure 6.55. (*) This data point is the result of the fit of the η-signal in the corresponding projection (see text). 157

6.18. Applied *time* cuts in the exclusive analysis of $\eta \to \pi^+\pi^-\pi^0$. 162

6.19. Applied *time* cuts in the exclusive analysis of $\omega \to \pi^0\gamma$. 164

6.20. Applied *time* cuts in the exclusive analysis of $\omega \to \pi^+\pi^-\pi^0$. 168

6.21. Applied *time* cuts in the exclusive analysis of $\pi^0\eta$-production. 170

6.22. Results of the investigation of the $\pi^0\eta$ production cross section. The acceptance for each interval is listed in Table 6.11. 172

6.23. π^0-Dalitz analysis: data points. 177

6.24. List of analyzed decays in the July beamtime, reconstructed events and the corresponding acceptances; the values of the branching ratios were taken from [20]. 178

6.25. Investigation of the contribution of the conversion effect to the background in the η-Dalitz analysis (see text). 183

7.1. Listed are the analyzed decay channels, number of reconstructed events, the acceptance (for an exclusive analysis) and the results. The listed information only refer to the data set obtained in the July beamtime 2007. 198

A.1. List of applied cuts in the η-Dalitz analysis. Based on a simulation of 2.5 million η-Dalitz events the relative strength of each cut was tested (see section 6.1.3). Since cuts are not independent the overall intensity reduction due to cuts is not the product of all factors. The detection efficiency for the η-Dalitz channel (after cuts) was determined as 1.3 %. . 218

A.2. Materials (and their radiation length) in the target region. The 4th column lists the probability for a γ to pass through the corresponding medium (without conversion). 218

A.3. Data points obtained in the η-Dalitz analysis. See section 6.24 (Fig. 6.56-6.58) and 7.3 (Fig. 7.4-7.6). 219

A.4. Applied cuts in the analysis of $\pi^0\pi^0$-events. 220

C. List of Tables

D. Bibliography

[1] M. N. Achasov et al., *Study of the conversion decays $\phi \to \eta e^+e^-$ and $\eta \to e^+e^-\gamma$ in the experiment with SND detector at VEPP-2M collider*; Physics Letters B (2001); B(504):275-281 25, 26

[2] J.R.M Annand, *The Glasgow/Mainz Bremsstrahlung Tagger Operations Manual*, 2008

[3] J.R.M Annand, *Data analysis within an AcquRoot Framework*, University of Glasgow, 2005

[4] R. Baldini et al., *The inverse problem: Extracting time-like form space-like data* (2001)

[5] L.M. Barkov et al., *Electromagnetic pion form factor in the time-like region*; Nucl. Phys. (1985): B(256):356-384 23

[6] H. Berghaeuser, *Untersuchung des η-Dalitz-Zerfalls und Bestimmung des η-Formfaktors mit CB/TAPS @ MAMI*, Diplomarbeit, Universität Giessen

[7] Website of Heninng Berghaeuser *http://www.henningberghaeuser.de*

[8] Yu. B. Bushinin et al., *Observation of the decay $\eta \to \mu^+\mu^-\gamma$*; Physics Letters B (1978); 79(1,2):147;doi:10.1016 / 0370-2693(78)90456-2-26

[9] E. Bosze, J. Simon-Gillo, J. Chang, J. Boissevain and R. Seto. *Rohacell foam as a silicon support strucutre material for the PHENIX multiplicity vertex detector*. Nuclear Instruments and Materials in Physics Research A, 400:224-232, 1997

[10] Y. Chan et al., *Design and Performance of a Modularized NaI(Tl) Detector*, IEEE Article, 1977

[11] B1 Collaboration, *http://www.kph.uni-mainz.de/B1/*, 2009

[12] S. Damjanovic et al., *NA60 PLB 677 (2009) 260*

[13] R. I. Djhelyadin et al., *Investigation of the electromagnetic structure of the η meson in the deacy $\eta \to \mu^+\mu^-\gamma$*; Physics Letters B (1980); 94(4):548; doi:10.1016 / 03702693 (80) 90937-5 26

[14] E.J. Downie, *Radiative π^0 photoproduction in the region of the $\triangle(1232)$ resonance*, Dissertation, University of Glasgow, 2006

D. Bibliography

[15] P. Drexler, *Entwicklung und Aufbau der neuen TAPS-Elektronik*. PhD thesis, Justus Liebig Universität Giessen, 2004

[16] EJ-204 Plastic Scintillator Data Sheet, *http://www.eljentechnology.com/datasheets/ EJ204%20data%20sheet.pdf*

[17] T. Geßler, *Particle Identification of Reaction Products from Photonuclear Reactions Using the TAPS Detector System*. Bachelor Thesis, Justus Liebig Universität Giessen

[18] D. Glazier, *A2 simulation handbook*, University of Edinburgh

[19] Particle Data Group, *Particle Physics Booklet; American Insitute of Physics, 2004*

[20] Particle Data Group *Particle Physics Booklet, July 2006*

[21] S. J. Hall, G. J., R. Beck, and P. Jennewein, *A focal plain system for the 855 MeV tagged photon spectrometer at MAMI-B*. Nuclear Instruments and Methods in Physics Research A, 368:698-708, 1996

[22] H. Herminghaus, *First operation of the 850 MeV c.w. electron accelerator MAMI*. In Proc. of the 1990 Linear Acc. Conf. Albuquerque, N.M, 10.9-14.9.90, page 362, 1990

[23] P.W. Higgs, *Broken Symmetries and the Masses of Gauge Bosons*. Phys. Rev. Lett. B 13 (1964) 508.

[24] M. R. Jane et al., *A measurement of the electromagnetic form factor of the eta meson and of the branching ration for the eta dalitz decay; Physics Letters B (1975); 59(1):103 doi:10.1016 / 0370 - 2693 (75) 90168-9 25*

[25] A. Jankowiak, *The Mainz Microtron MAMI - past and future*. Technical Report 1, KPH, 2005

[26] V.L. Kashevarov et al, *Photoproduction of $\pi^0 \eta$ on protons and the $\Delta(1700)D33$ resonance,EPJ A 42 (2009) 141, DOI 10.1140/epja/i2009-10868-4, 9 Pages: 141-149*

[27] K. Kleinknecht, *Detektoren für Teilchenstrahlung*, Teubner , 2005

[28] S. Klimt, M. Lutz, W. Weise, *Chiral phase transition in the SU(3) Nambu and Jona-Lasinio model; Phys. Lett. B (1990); 249:386 18*

[29] D. Krambrich, *Aufbau des Crystal Ball-Detecktorsystems und Untersuchung der Helizitätsasymmetrie in $\gamma p \to p\pi^0\pi^0$; Doktorarbeit, Universität Mainz, 2007*

[30] B. Krusche, *Photoproduction of π^0 and η mesons from nucleons und nuclei in the second resonance region*. Habilitation thesis, Justus-Liebig-Universität Gießen, Germany, 1995

[31] L.G. Landsberg, *Electromagnetic decays of light mesons; prep (1985); 128:301 21, 22, 25, 26, 98*

[32] B. Lemmer, *Measurement of the Excitation Function of ω Photoproduction an Carbon and Niobium*, Diploma Thesis, University of Giessen, 2009

[33] E. Lohrmann, *Hochenergiephysik*, Lehrbuch 5. Auflage, Teubner

[34] S. Lugert, *In-Medium Modification of Pion-Pairs on Deuterium*, Dissertation, Universität Giessen, 2007

[35] M. F. M. Lutz, S. Leupold, *A 813 (2008) 96-170*

[36] B.M.K Nefkens, *The Crsytal Ball - Overview; Crystal Ball Report 95-1*, 1995

[37] A. Nikolaev, *PhD thesis, University of Mainz*, 2007

[38] R. Novotny, *The BaF_2 spectrometer TAPS*. IEEE Transactions on Nuclear Science, 38:379-385, 1991

[39] R. Novotny, R. Beck, W. Döring, V. Hejny, A. Hofstaetter, V. Metag and K. Römer. *Scintillators for photon detection at medium energies - a comparative study of BaF_2, CeF_3 and $PbWO_4$*. Nuclear Instruments and Methods in Physics Research A, 486:131-135, 2002.

[40] R. Novotny. *The BaF_2 spectrometer taps: A system for high energy photon and neutral meson detection*. International Journal of Radiation Applications and Instrumentation. Part D. Nuclear Tracks and Radiation MEasurements 21(1):23-26, 1993

[41] N. P. Siamos, *Dynamics of Internally Converted Electron-Positron Pairs*, Physics Review, 121 January 1, 1961

[42] C. F. Redmer, *In search of the Box-Anomaly with the WASA facility at COSY*, Dissertation, Bergische Universität Wuppertal.

[43] A. Reiter et al., *A microscope for the Glasgow photon tagging spectrometer in Mainz*. European Physical Journal A, 30, 2006

[44] Povh, Rith, Sholz, Zetsche *Teilchen und Kerne*, Springer Verlag

[45] M. Röbig, *Eichung des TAPS-Detektorsystems mit Höhenstrahlung*; Universität Giessen, 1991

[46] C. Terschlüsen, *Elektromagnetische Übergangsformfaktoren für pseudoskalare und Vektormesonen*, Diplomarbeit, Universität Giessen, 2010

[47] M. Thiel, *In-medium modifications of the ω-meson*, Dissertation, University of Giessen, 2011

[48] U. Thoma, *Private Communications*, thoma@iskp.uni-bonn.de

[49] A. Thomas, *Crystal Ball Hydrogen (Deuterium) Target Manual*; Vortrag, Universität Mainz, 24.06.2004

D. Bibliography

[50] M. Unverzagt, *Energieeichung des Crystal Ball Detektors am MAMI*, Diplomarbeit , Universität Mainz, 2004

[51] S. Wartenberg, *Die Strahlungsasymmetrie in der Deuteron-Photospaltung im Bereich von 160 bis 410 MeV;* Universität Mainz, 1997

[52] http://commons.wikimedia.org/wiki/File:Baryon-octet-small.svg

[53] http://commons.wikimedia.org/wiki/File:Baryon-decuplet-small.svg

Danksagung

Mein Dank gilt an erster Stelle Herrn Prof. Dr. Volker Metag für die Bereitstellung der Dissertation in diesem interessanten Themengebiet der Physik und für die bereitwillige Unterstützung bei der Durchführung und Auswertung der Arbeit sowie für die Möglichkeit der Mitarbeit in seiner Arbeitsgruppe.

Des Weiteren möchte ich mich bei den Mitarbeitern des 2. Physikalischen Institutes für die angenehme Arbeitsatmosphäre und das freundliche Miteinander bedanken, ganz besonders hierbei bei Frau Michaela Thiel, meiner Bürokollegin.

Furthermore I would like to express my thanks to several members of the A2-Collaboration. Without Dominik Werthmüller from the University of Basel I would have had certainly more difficulties in getting started with the calibration of the data. His macros provided a perfect basis for my own developments. Manuel Dieterle, University of Basel, accomplished all the converter work concerning the performed simulations - thank you. I would like to give special thanks to Dr. J.R.M. Annand, University of Edinburgh: AcquRoot is really a nice program and I like it. Moreover I want to thank Dr. Sven Schumann and Dr. Evangeline Downie (University of Mainz) for their advisory support (concerning experimental & programming issues).

Mein ganz besonderer Dank gilt meinen Eltern und meiner lieben Lebensgefährtin Inga Skileva für die stete Unterstützung und Geduld und Verständnis, die mir nicht zuletzt den erfolgreichen Abschluss meiner Arbeit erleichterten.

I want morebooks!

Buy your books fast and straightforward online - at one of world's fastest growing online book stores! Environmentally sound due to Print-on-Demand technologies.

Buy your books online at
www.morebooks.shop

Kaufen Sie Ihre Bücher schnell und unkompliziert online – auf einer der am schnellsten wachsenden Buchhandelsplattformen weltweit! Dank Print-On-Demand umwelt- und ressourcenschonend produziert.

Bücher schneller online kaufen
www.morebooks.shop

KS OmniScriptum Publishing
Brivibas gatve 197
LV-1039 Riga, Latvia
Telefax +371 686 204 55

info@omniscriptum.com
www.omniscriptum.com

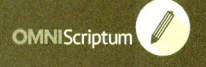

Printed by Books on Demand GmbH, Norderstedt / Germany